S0-ABA-217

Risk-Based Management

A Reliability-Centered Approach

Risk-Based Management

A Reliability-Centered Approach

Richard B. Jones

Gulf Publishing Company
Houston, London, Paris, Zurich, Tokyo

To Janice, Jaime, Ricky, and Indy

Copyright © 1995 by Gulf Publishing Company, Houston, Texas. All rights reserved. Printed in the United States of America. This book, or parts thereof, may not be reproduced in any form without permission of the publishers.

Gulf Publishing Company
Book Division
P.O. Box 2608 ☐ Houston, Texas 77252-2608

Printed on Acid-Free Paper (∞)

10 9 8 7 6 5 4 3 2 1

Library of Congress Cataloging-in-Publication Data

Jones, Richard B. (Richard Bradley), 1947–
    Risk-based management : a reliability centered approach / Richard B. Jones.
        p.   cm.
    Includes index.
    ISBN 0-88415-785-7
    1. Plant maintenance. 2. Plant maintenance—Management. 3. Reliability (Engineering)   I. Title.
TS192.J66   1995
658.2'02—dc20   94-36659
                                                CIP

Contents

v

vi

Acknowledgments

First and foremost, I owe the existence of this book to Janice Jones, whose professional editorial expertise and unique ability to transform my words and thoughts into plain English were invaluable. Janice's talent for writing, keeping this humble applied mathematician grounded in reality, and overall attention to detail are what made the creation of this book possible.

My special thanks to Maurice Darbyshire and Bill Voegtle for countless discussions on the practical aspects of risk modification and for providing reality checks by asking, "What exactly are you trying to achieve?" I am also grateful to Ray Oliverson and Bernie Selig for the many helpful conversations we had and for their help in improving the manuscript with their comments and criticisms. I am also indebted to Jim Sutherlin for his overall encouragement and, more specifically, for providing a business perspective during the long process of consolidating and distilling the seemingly innumerable facts and figures into manuscript form.

The material compiled for this book required extensive literature searches, and I would like to thank Peter Moon for his valuable help. Peter always kept me in mind when new titles came to his attention and even sent me papers that he thought I'd be interested in. He was always right!

A large part of the material in this book reflects countless discussions with plant personnel at locations all over North America. There are too many people to explicitly name, but these mostly informal conversations formed the overall philosophy that is developed in the book. These people are the day-to-day practitioners of reliability improvement, risk reduction, and risk-based management. To each of them, I say "Thank You!"

Preface

My studies in the theoretical world of reliability once gave me what I thought were powerful analytical tools for solving real-world problems with accuracy and precision. The elegant logic engines developed around the pure rigors of mathematics, probability, and statistics were extremely attractive to an applied mathematician. Unfortunately, my little bubble of numerical security popped every time I set foot in a chemical plant, refinery, discrete manufacturer, paper mill, and, in fact, basically any type of real-life situation. The reliability papers I had relied on were built largely with a mathematical elegance that exists primarily in the rarefied air of theory. Although they offered tremendous intellectual challenge and have advanced the state of the art in reliability theory, they often missed the mark where "the rubber meets the road." I found that, in most cases, the models I'd embraced were running on empty. That is, they were built with little real data, using assumptions that could not be verified. Most maintenance and operations people were inclined to look at them, as I came to, as houses of straw. In real life, prospective beneficiaries of the ideas were likely to read the papers and say, "So what? Nice ideas, but what am I supposed to do with them? How are they going to improve the bottom line at *my* plant?" These people do not want theory or untested methods. They want answers. They want guaranteed results, quantifiable reliability improvement at prices they can afford. This is a much greater challenge. It involves people, their attitudes, their perceptions, and their cultures, as well as the performance of their equipment.

This book narrows the gap between theory and actual plant operations and maintenance. It supplies information and ideas that consider the data quality and quantity limitations that exist. It is written to provide a foundation and a direction for gleaning information from an imperfect world. There are two practical constraints on the analytical topics included in the book. They are:

1. All mathematical procedures must use only data available in common industrial situations.
2. The results must be directly useful to plant personnel and help them ultimately manage and improve reliability and reduce risk.

The product is what I call "Risk-Based Management." Risk-based management is a complex structure that requires the application of several principles, just as a multi-dimensional object requires viewing from several dimensions. This book is a resource for ideas and methods that have been used to improve reliability and measurement, and reduce risk in a wide variety of industrial situations. It discusses the major principles of dimensions of industrial risk-based management. There is no one, simple recipe for risk-based management. My hope is that the book will help you develop the proper mix of risk management techniques for your particular company, plant, or business.

The title of the book includes the phrase "A Reliability-Centered Approach" because risk-based management is built on the principles of reliability-centered maintenance (RCM). The fundamental premise of RCM is to maintain system function by analyzing how a system functions and how it can functionally fail. I believe this philosophy is not restricted in applicability to just maintenance organizations; it is applicable to many businesses, including service organizations. However, before its principles can be extended they must be firmly understood. Thus, a portion of the book is directed towards understanding RCM, describing current industrial RCM experience, and discussing how to incorporate risk into RCM in a realistic and technically meaningful way.

From my perspective, risk reduction is an industrial-strength version of "continuous improvement." Although we will never achieve a zero risk activity, risk-based management principles supply tools to help us make decisions in this competitive, regulated, and imperfect world. Eliminating risk altogether may not be an achievable goal, but I believe that risk reduction is an essential part of every corporation's management portfolio. I sincerely hope that after reading this book, you will agree.

Richard B. Jones, Ph.D.

CHAPTER 1

Evolution of Reliability-Centered Maintenance

"There are risks and costs to a program of action, but they are far less than the long-range risks and costs of comfortable inaction."

—*John F. Kennedy*

What is Reliability-Centered Maintenance?

Reliability-centered maintenance (RCM) was created to provide guidelines on which equipment should be addressed by which maintenance tasks and at what frequencies. It is not a mathematical formula. It is a method by which a company can use its failure data, system design redundancies, and operating experiences to develop a flexible and cohesive maintenance design. Before I go any further, let me formally define reliability-centered maintenance.

> **Reliability-Centered Maintenance is:**
> A method for developing and selecting maintenance design alternatives, based on safety, operational, and economic criteria. RCM employs a system perspective in its analyses of system functions, failures of the functions, and prevention of these failures.

This definition of reliability-centered maintenance is intended to apply to all variations of the method. It represents the overall philosophy of RCM more than a specific means of implementation.

Before we get into the *hows* of RCM, let's see *why* it came to be.

Change reaches us in one of two ways; it can be evolutionary, modifying circumstances and methods over a period of time, or more often in today's fast-paced business environment, change comes as a quick substi-

1

tution of one direction for another. Even the most tried and true techniques and methods may eventually become obsolete simply because better ways of solving problems are found. Changes come in many forms, but in industry are usually intended to improve profit, safety, compliance with regulations, and even the ability of the industry to survive.

The changes defined by RCM have evolved over recent years. RCM was developed by, and for, the United States commercial airline industry in cooperation with the federal government, aircraft manufacturers, and air carriers. Since 1974, the U.S. Department of Defense has used RCM as a maintenance philosophy for military aircraft systems. The commercial nuclear power industry has adopted the method and is implementing its development across the enterprise. There are many other industries that have implemented RCM, and even more are actively studying the RCM philosophy to decide exactly how it will fit into their environments, industrial systems' designs, and their cultures.

To best understand RCM and the concerns of the airline industry that forced its development, and how RCM was created and has evolved, we'll begin with a look at the early days of airline operation.

The Beginnings of Maintenance Policy: Equipment Maintenance

The McDonnell Douglas DC-3 was a reliable aircraft and economical to operate. It was a simple design by today's standards, with a straightforward assembly procedure requiring no highly specialized tools. The equipment and expertise of the personnel required to maintain the aircraft were largely satisfied by existing airline staffs.

During the 1930s, this aircraft was being licensed for commercial use. At this point, the airline industry was in its infancy and had minimal information concerning the age, reliability, and safety relationships of the DC-3's components and systems. Industry experts had little knowledge and experience to help them determine which maintenance tasks should be done and how frequently each should be performed. Because public safety was the major concern of the fledgling industry, an extremely conservative maintenance policy was created, based on frequent disassembly and inspection of each aircraft [1]. Inspection criteria were defined without substantive assurance of the need for them and with no proof that they enhanced safety. The airframe was stripped and inspected every three years [2]. These detailed, periodic inspections and accompanying part replacements provided the original knowledge base for determining aircraft reliability.

Expansion in the Airline Industry: Condition Monitoring and Simulation

The next twenty to thirty years brought rapid advancement in aircraft equipment and systems. These improvements permitted greatly increased speed, capacity, and versatility in distance and altitude for the aircraft. They allowed the airline industry to become a vital part of life in the 20th century. Yet in spite of the tremendous progress in technology and resulting capabilities, and the growing importance and economic value of the airline industry, little improvement came in the methods that were used for maintenance.

One of the first maintenance advancements was made during the early 1950s, when condition monitoring equipment was added to aircraft systems. This equipment provided continuous information about equipment function and detected problems in the equipment as they occurred. It allowed maintenance to be performed based on need as well as on schedule.

Also during this time, laboratory testing was developed to simulate the life cycles of aircraft parts. Results of this type of testing were used to help determine the schedules for disassembly and inspection of the actual components.

Throughout this time period, however, this basic premise remained— equipment was tested and maintained on a unit/type basis, with little or no regard for the relationships of each component with its counterparts and the environment. Each serviceable item was directly targeted by a standalone maintenance policy.

MSG-1: A Maintenance Breakthrough

When the Boeing 747 and Lockheed L1011 aircraft were introduced in 1968, the concern for public safety took on a new dimension. For the first time in history, over 300 members of the general public could be flown in a single airplane. The complexity, sheer size, and number of equipment units of these aircraft challenged the aircraft industry and the Federal Aviation Administration (FAA) to develop a logical process and design appropriate maintenance tasks with optimal frequencies for these tasks. These complex systems absolutely could not be maintained using policies that originated in the relatively simple DC-3 era.

In 1968, representatives of various airlines, aircraft manufacturers, and the U.S. government organized a Maintenance Steering Group (MSG-1). Its task was to establish the appropriate maintenance procedures for the

B-747 and L1011. This group specifically sought to reduce maintenance downtime, reduce maintenance costs, and improve flight safety in one fell swoop. As a result, *MSG-1: Maintenance Evaluation and Program Development* [3] was published. A Lockheed official described the document by saying, "These guidelines provided the first formalized breakthrough in establishing new criteria for maintenance programs. They replaced maintenance concepts that had been in use for almost 60 years." [4]

In more contemporary terms, it can be said that MSG-1 reengineered airline maintenance.

As we discussed, prior to MSG-1, maintenance was performed on each unit of equipment without considering the importance of that unit to system function. The MSG-1 study changed the maintenance paradigm. It recognized that aircraft are highly redundant, complex systems. The orientation changed from examination of *equipment function* to that of *system function.*

Here is an example of the differences between an equipment perspective and a systems perspective in maintenance. Let's suppose that there are eight hydraulic control valves on an airplane. To lose hydraulic control functions, all eight of these valves must fail. The manufacturer of the valves recommends that the valves be replaced every 2,000 hours of flight time. If an equipment-oriented maintenance policy is followed, all eight valves would be replaced every 2,000 hours, even though it is likely that they remain fully functional. A system function maintenance perspective, however, recognizes the fact that 2,000 hours is an *average* lifetime for each valve, and not an absolute limit on safety. It considers the fact that the probability of all eight valves failing at the same time is infinitesimally small. The system-oriented maintenance policy incorporates the inherent safety redundancy of the valves into its formulation. Only one valve is replaced every 2,000 hours by systematically selecting a different valve location each time until all eight locations have been changed. The process is then started over with the first valve. As a result, there will always be at least one valve with fewer than 2,000 hours of use. This saves time and money and has proven a highly effective and safe strategy. These maintenance decisions are based on more complete and accurate system information. It is important to remember that they are still made by the people who know the systems.

Airline/Manufacturer Maintenance Program Planning Document (MSG-2) [5]

The goals of MSG-1 were clearly not unique to the B-747 and L1011. Maintenance costs, complexity, and safety concerns were widespread. In 1970, a second Maintenance Steering Group (MSG-2) was formed. This committee generalized the specific maintenance procedures of MSG-1 to make them applicable to other aircraft. Their work incorporated a very powerful, yet simple, tool called decision tree logic. A decision tree is a diagram that provides a logical, hierarchical sequence of questions about a series of possible events and their outcomes. Each question in the decision tree can be answered only by YES or NO. The outcome of each question leads either to an action to be taken or to the next question in the series. Figure 1-1 contains an example of a decision tree.

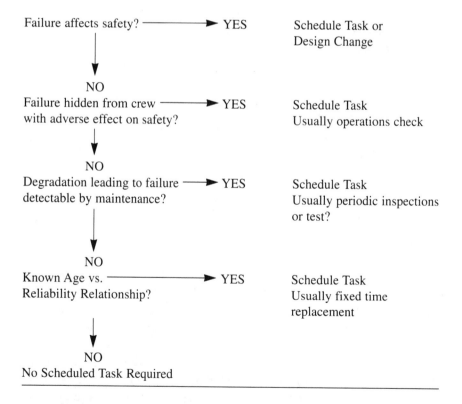

Figure 1-1. MSG-2 Decision Tree Logic

The tree acts like a logical road map. Each possible system failure is categorized by applying the tree logic of YES and NO questions. It directs the examiner through a logical analysis that ends when a YES answer is obtained. For each NO answer, the examiner proceeds to the question below. If the bottom of the tree is reached, then the logical conclusion is that no task is required for the failure under examination.

The results of MSG-2 were published by the Air Transport Association (ATA). The FAA reviewed the document and approved it as a reasonable, practical guideline for establishing new aircraft maintenance requirements [6]. MSG-2 is also known as reliability-centered maintenance. It soon became a standard for maintenance design of aircraft in the planning, development, and operational phases.

The success of RCM in the airline industry was unprecedented. *Over the 16-year period following its implementation, the airlines had no increase in maintenance costs per unit even though aircraft size and complexity and the costs of labor increased greatly during this time. Over the same period, the airlines' safety records improved as well* [7]. These achievements were due to a number of factors including improvements in equipment, more experience, and the adoption of RCM for aircraft maintenance. There are few, if any, industries that can match this reengineering performance record.

In 1980, the knowledge accumulated during the last half of the 1970s was used to update MSG-2 procedures. This work was performed by a group similar to the Maintenance Steering Group-2, and was published as the Airline/Manufacturer Maintenance Program Planning Document (MSG-3) [8].

Applications of RCM

RCM in the Military

The benefits of RCM enjoyed by the airlines and the FAA were no secret. The Department of Defense was also under pressure to reduce operations and maintenance costs without sacrificing its overall readiness capability. In the early 1970s, Congress, in cooperation with the Office of the Secretary of Defense, initiated a study to examine RCM applications in the U.S. military. There were many attractive features of RCM that appeared to have direct application in the military environment. However, some military officials argued that the mission of military aircraft was radically different from that of commercial airline aircraft. What works well for airlines, they theorized, could have little application to the military world. Congressional pressure to reduce aircraft overhaul costs won out over the "religious wars" of the military officials and forced a serious attempt to apply RCM to military aircraft. There were two reasons why airline experience with RCM was convincing:

1. The reliability decision tree logic approach leads to greater effectiveness in providing a standardized justification process for doing and *not doing* maintenance.
2. By reducing the amount of time an aircraft spends in the shop, more operational time can be enjoyed, thus reducing the need to increase the number of aircraft committed to each individual mission.

In 1974, the Department of Defense adopted the RCM approach as a basis for maintenance of military aircraft systems [9–10]. During the second half of the 1970s, RCM was integrated into all of the U.S. military maintenance programs.

Military results have been successful in either maintaining or increasing availability while enjoying formidable cost savings and cost avoidance. The method has significantly reduced the number, frequency, and content of overhauls. This reduced depot activity and provided cost savings in inventory levels and other related areas.

RCM in the Nuclear Industry

In the early 1980s, the US. commercial nuclear power industry began to study the application of RCM to their highly specialized systems. The question was: *Is a maintenance management method successful for civilian and military aircraft systems also suitable for a nuclear power plant?*

From a systems perspective, all are complex, redundant entities that have a high degree of inherent reliability. From a regulatory perspective, all have government agencies monitoring their operation. (The airlines have the FAA, the military has Congress, and nuclear power utilities have the Nuclear Regulatory Commission.) The similarities were encouraging. In 1983, the Electric Power Research Institute (EPRI), the research and development arm of the utility community, began RCM feasibility studies at selected reactor sites.

Within two years, reports of nuclear power plant RCM applications showed promising results. The initial study describes an RCM application for a cooling water system [11]. This system operates under all power plant conditions including startup, power production, emergency shutdown, and normal shutdown. It is used to circulate cooling water needed for the operation of both safety-related and power production equipment. From an equipment complexity perspective, the system contains a variety of pumps, pipes, valves, heat exchangers, and instrumentation-control segments. It is a fairly complex system. When the study was completed, the RCM derived maintenance design actually had *more* tasks than the original maintenance program [12]. However, the type and frequency of tasks changed considerably. Several labor and material intensive tasks that were performed at specified time intervals before the RCM feasibility study began now were performed only when the equipment degraded to certain measurable conditions. The overall savings was expected to be a 30 to 40% reduction in maintenance labor-hours and material costs. Additional savings were expected due to a reduction in forced outages from system failures.

While RCM studies were taking place for nuclear applications, uses of RCM were becoming more commonplace elsewhere also. They were especially prevalent in Europe and Australia at this time. RCM was being used together with other techniques such as Reliability, Maintainability, and Availability Analysis to develop maintenance designs. Industries started to become more varied from the standard aviation, military, or nuclear power utilities. Here are some examples.

Maritime Applications of RCM

Maritime applications of RCM were demonstrated using a 75-ton chiller unit on board a destroyer [13]. The unit consisted of six subsystems: compressor, controller, 75-ton chiller unit, 75-hp motor, condenser/cooler, and shell package.

The chiller system was considered highly reliable. This study sought to improve its maintenance plan by addressing the following concerns:

- Which components required maintenance?
- What were the optimal maintenance intervals for major overhauls (refits) of the equipment?

This project coupled RCM with life-cycle costing principles. Component maintenance tasks were determined using RCM. "Optimal" overhaul intervals were determined using life-cycle costing analysis. The longer the time between refits, the more preventive maintenance (PM) needed to be performed. The results of the study indicated only the compressor subsystem needed preventive maintenance during the refit life cycle. In addition, considerable cost savings were realized by reducing the number of hours in performing the refit. It was estimated that the refit time could be reduced from the previous 800 labor-hours to 300 labor-hours per unit.

This example shows how RCM techniques can be used in concert with economic constraints to improve the efficiency of overhaul intervals and overhaul tasks. Overhaul tasks and frequencies are related to the routine tasks and frequencies. If the PM intervals are very small, it may be possible to extend the overhaul intervals. However, when an overhaul is finally performed, it may also take more time and more money than if the interval from the last overhaul was smaller. On the other hand, if no PM is performed at all, the overhaul interval would be relatively short. The RCM method helped determine what PM should be done and how often.

RCM in a Solar Receiving Plant

Solar Central Receiver Plants [14] are designed to convert solar energy into electricity on a large scale. An existing experimental facility was used to study the viability of RCM for this power generation method.

The facility consisted of a single, tower-mounted receiver that served as the focus point for many parabolic dish sun reflectors. These reflectors could move to continuously direct the sun's energy to the central receiver. At the central receiver, the sunlight was concentrated onto receiver tubes. The circulating fluid in these tubes absorbed the thermal energy in the sunlight and transported the heat ($1,500°C$) to the conversion system where the heat was used in a steam boiler. A control subsystem allowed continuous regulation of flow, and a draining and venting subsystem allowed proper circulating fluid care and maintenance. The details of the receiver system used in this RCM study are shown in Table 1-1.

Results showed that the same methodology applied so successfully to the airlines and the nuclear power industry could also be applied to this type of system. The major steps of the method were applied without modification from previous applications. The primary conclusion of the study

Table 1-1
Solar Plant Receiver System Composition

Distribution	Heat Transfer	Control	Drain/Vent
Piping	Piping	Piping	Piping
Headers	Preheat Boiler Panels	Flow Control Valves	Valves
Manifolds	Superheat Boiler Panels	Sensors	Fluid
Filters	Moisture Separators	Flash Tank	

was that the RCM method could be applied without modification to non-aviation and non-nuclear systems. Feasibility was established.

RCM in Grain Terminals

At first glance, most people would consider a grain terminal too simple a system for the RCM approach. It's true that there are not as many functional redundancies in its systems as in previously mentioned industries. However, the grain terminal possesses subsystems that lend themselves to the RCM approach coupled with other reliability analysis techniques. In this case, a new grain terminal was being constructed and a maintenance plan had to be developed before the unit could be commissioned into operation. Here is a short description of the equipment contained in the terminal and the different functions that are performed by its various systems.

The grain terminal receives grain from rail and road stations. This particular terminal received the majority of tonnage via rail. At transfer sites, grain is discharged into a hopper where the bulk, temporary storage is contained until it can be transported via a variable-speed conveyer to grain storage into the terminal. Because grain amounts are measured in weights, weighing is an important function. During transit to the terminal, grain amounts are measured using a belt weigher. Automatic grain sampling is also performed in transit to examine the grain type and quality.

At the terminal, the grain is elevated to silo storage by a series of conveyor belts. The silos are freestanding and are gas-tight to resist grain contamination by insects and other agents. If unacceptable quality or infested grain is identified in transit to the silo, the shipment can be diverted to a pre-assigned silo or the inspection can result in the fumigation of the entire silo.

Dust is a major potential safety and reliability problem. All conveyor transfer points have dust collection systems and bag filters. Dust accumulation at the rail and road grain entry sites is mitigated by baffles. Both fixed and mobile vacuum units are used to keep airborne dust levels to a minimum, safe level. The collection system transports dust to a central collection repository that requires servicing on a periodic basis.

The last major function of the grain terminal is to transport stored, insect-free, high-quality grain to ships and other bulk carriers. Grain is transported to the loading site via conveyors. The grain enters a telescopic downspout and is deposited into the bulk carrier. Dust collection systems operate during the loading process to reduce airborne losses, employee health, and explosion potential.

The development of a maintenance design incorporating task contents and tasks frequencies took 14 weeks. Using the basic concepts of RCM helped the company determine which equipment and components were important and which were not. The result was an efficient, systematic procedure that helped people knowledgeable with grain terminal operations make the maintenance design decisions [15].

RCM in Coal Mining

In the coal mining industry, maintenance costs were not historically a major factor in business management. During this time period, though, the industry began to look for ways to cut production costs (including maintenance) and simultaneously increase equipment reliability. Coal mining is an equipment-intensive industry and even small equipment reliability increases could produce considerable savings in production budgets.

Coal mining maintenance historically has similarities with many other industries. Each piece of equipment had a maintenance program that was developed based on manufacturers' recommendations. These programs were based on tasks and frequencies used for previous or similar equipment. Certain maintenance activities had become standardized and were basically unchanged even though significant changes in equipment technology had occurred. As a result, there were maintenance tasks being performed with no justification and there were new failure modes not being addressed by the maintenance program. Because of the almost constantly changing equipment, management's energy and interest were focused on the equipment operation and effects on production. Maintenance redesign was not given a high priority.

The result of this policy in the mining industry was a fairly rapid increase in maintenance costs. It was not uncommon to have maintenance costs as high as 50–60% of the total equipment operating cost [16]. Occasionally, maintenance costs would contribute up to 30% of the total production cost [17]. These expenses were significant enough to justify a policy change for coal mining equipment maintenance.

As a result, the RCM method, coupled with reliability modeling techniques, were applied to a large haul dump machine [18]. The purpose was to conceptually: examine the use of RCM to determine what maintenance tasks to perform, and demonstrate a simple probabilistic approach for determining optimal task intervals.

The large haul dump machine results demonstrated the clear potential for the overall applicability of RCM and also showed how standard reliability methods could be used in concert with RCM. The cost effectiveness in coal mining operations also appeared to be significant. This initial work demonstrated feasibility of the method. It is significant that the RCM concept can be applied to a B-747, F-18, and also a coal mine large haul dump machine.

Applications have been reported in oil refineries [19] and gas plants [20]. Paper mill applications have also been performed. The RCM method developed for the commercial airlines has been adapted to fit the needs of a wide range of diverse industries. The basic principle of *Maintain System Function* is always recognizable even though some of the details have been changed. The method is still evolving as it is adapted by creative people to the changing needs of a wide range of industries. The versatility of the method and the method's basic principle give credence to the postulate that failures are indeed not random and are driven by the basic laws of physics.

There are many reasons why a "system function" approach to maintenance rather than an "equipment based" approach is a better paradigm for increasing reliability and safety while decreasing maintenance costs. No book can attempt to construct an exhaustive list. There is, however, one subtle but *very important* reason that I would like to point out. Complex, redundant systems have reliability directly engineered into their design. The reliability of the system can be reduced by maintenance if the tasks and frequencies are not a part of the design. Too much maintenance can reduce system reliability from maintenance-induced failures. *For highly reliable systems, the most likely failure can be human intervention under the pretense of preventive maintenance.* An Air Force study on this subject found that 40% of the work required to restore a sample of F-4 Phantom jets to operational condition was the direct result of failures induced by previous maintenance [21].

Then Why Don't We All Use RCM?

Improvement activities are not undertaken without profit motivation. Belts are being tightened; downsizing and reorganization are common activities in many corporations, both large and small. Today, survival, not just prosperity, is the concern of many companies. Increasing taxation, changes and uncertainties in government regulations, and the public demand for perfection from industries make *status quo* maintenance policies a luxury few, if any, equipment-intensive businesses can afford. RCM is a proven method for optimizing the use of maintenance personnel, time, and money. It is an idea whose time has come.

This book discusses reliability-centered maintenance and risk from a point of view that enables practitioners to apply these concepts to their business or industry. *You* are challenged to relate the concepts, issues, and examples of this book to your plant, industry, and culture. Applying RCM and risk measures is not a "leap of faith" but a tried and proven program for action. Global competition and innumerable other factors are making any program of "comfortable inaction" a very high risk activity.

References

1. Connell, F. H., "P-3 Improved Maintenance Program," *Lockheed ORION Service Digest,* Lockheed California Co., March 1976, p. 9.
2. Viereck, E. A., Jr., Col. USA, "Reliability-Centered Maintenance Strategy for U.S. Army Equipment," *Proceedings: Maintainability Engineering Symposium,* Orlando, Florida, 1977 p. 56.
3. *747 Maintenance Steering Group Handbook; Maintenance Evaluation and Program Development (MSG-1),* Air Transport Association, July 10, 1968.
4. Connell, F. H., "P-3 Improved Maintenance Program," *Lockheed ORION Service Digest,* Lockheed California Co., March 1976, p. 10.
5. *Airline/Manufacturer Maintenance Program Planning Document: MSG-2,* R & M Subcommittee, Air Transport Association of America, Washington, DC March 25, 1970, p. 1.
6. Ref. 4, p. 11.
7. Gaertner, J. P. and Worledge, D. H., "Prospects for Reliability-Centered Maintenance," American Power Conference, April 1980, p. 598.
8. *Airline/Manufacturer Maintenance Program Planning Document: MSG-3,* R & M Subcommittee, Air Transport Association of America, Washington, DC October 1980, p. 1.
9. Saia, J. C., Civ. USN, "An Evaluation of Changes in the U.S. Navy Aircraft Maintenance Program," Master's Thesis, American University, Washington, DC, March 1977, p. 48.
10. A'Hearn, F. W.,"The MSG-2 Commercial Airline Concept: A Department of Defense Program for Reliability-Centered Maintenance," Defense Systems Management College, Ft. Belvoir, VA PMC 77-2 (1977), p. 38.
11. "Application of Reliability-Centered Maintenance to Component Cooling-Water System at Turkey Point Units 3 and 4," EPRI NP-4271, October 1985.
12. Smith, A. M., Vasudevan, R. V., Matteson, T. D. and Gaertner, J. P., "Enhancing Plant Preventive Maintenance Via RCM," ARM (Annual Reliability and Maintainability Symposium) 1986, pp. 122.

13. Jambulingam, N. and Jardine, A. K. S., "Life Cycle Costing Considerations in Reliability-Centered Maintenance: An Application to Maritime Equipment," *Reliability Engineering,* 15, 1986, pp. 307–317.
14. Matteson, T. D. and Smith, A. M., "Application of Reliability-Centered Maintenance to Solar Central Receiver Plants," SAND-86-8177, 1986.
15. Carlton, A. J., and Lenox, D. E., "A Low Cost Maintenance Management System for a Materials Handling Plant," Maintenance Engineering Conference, Melbourne, Australia, May 1987, pp. 50–57.
16. Lindquist, P. A., "Mining Methods 2000" (in Swedish), Stiftelsen Bergteknisk Forskning, Befo, Stockholm, 1987.
17. Cutts, A. and Ford, D. C., "Monitoring the Reliability of Coal Face Equipment," Mining Conference on Reliability and Monitoring, United Kingdom, 1985, pp. 293–305.
18. Kumar, U. and Granholm, S., "Reliability-Centered Maintenance for Mining Equipment," 21st Century Production Coal Mining Systems Symposium, Victoria, Australia, 1988, pp. 343–349.
19. Jones, R. B., "An Application of Reliability-Centered Maintenance," First International Conference on Improving Reliability in Petroleum Refineries and Chemical and Natural Gas Plants, Houston, TX, Gulf Publishing Co. November 1992.
20. Pradham, S., "Application of Reliability-Centered Maintenance to the Management of Sealless Pumps in Critical Hazardous Services," First International Conference on Improving Reliability in Petroleum Refineries and Chemical and Natural Gas Plants, Houston, TX, Gulf Publishing Co. November 1992.
21. Smith, C., Civ. OASD (MRA&L), "Why and What is Reliability-Centered Maintenance?" unpublished paper, May 12, 1976, p. 5.

CHAPTER 2

A Measurement Perspective

From the intrinsic evidence of his creation, the Great
Architect of the Universe now begins to appear as a pure
mathematician.

Mysterious Universe (1930)
Sir James Jeans (1877-1946)

There is no such thing as a random event. Nature has given us a highly ordered world whose controlling processes are described by the fundamental laws of physics. This means that absolute randomness cannot exist in either man-made systems or the systems of nature. Unexpected occurrences in natural, mechanical and electronic systems, and business processes can generally be termed "failures." Failures are not spontaneous occurrences but are always preceded by an explicit sequence of events. When we do not detect a preliminary pattern of events and failures occur without apparent warning, we generally term them "random." In reality, a random event is merely one for which the metrics to measure its precursors are not in place, and in fact, are often not understood. Events become predictable when the appropriate measures and models to use the measurements are defined and employed. For example, many people view the weather on a given day as random. However, today's use of technology and forecasting systems to obtain and use meteorological measurements have begun to unveil the patterns and events leading to each day's weather. This, in turn, allows us to better predict and prepare for it. As this example shows, when effective measurements are defined and applied to a system, our perception of randomness is replaced with an understanding of system order.

Meteorology is just one of the many fields in which technology can and has revolutionized the types of procedures and precision involved in obtaining data. Today's wide variety of system and equipment monitoring and diagnostic devices can use the power of computer chips, graphics, and

15

software tools to track virtually every process, every transaction, and every system pulse with great accuracy. Unfortunately, our tremendous ability to monitor events is not always accompanied by equally strong improvements in planning for them, and more importantly, making changes to avoid the negative deviations from the status quo and encourage the positive ones. Rather than a great revolution in planning and preparedness, we're often left with a glut of unused data. Our challenge is to closely examine our systems and our needs, measure only the essential data, and then put our measurements to productive and profitable use.

A Measurement Experiment

Here is a simple measurement exercise that can illustrate important aspects of measurement and provide a good basis for discussion of measurement and error. You can conduct it as follows:

1. Assemble a small group of participants. Do not tell the group anything about the problem.
2. Outside of your meeting room, place a board, a tape measure or ruler, and instructions which read:
   Please use this ruler to measure the length of this board. Write your result on a piece of paper and give it to me as you reenter the room. Do not discuss your measurement result or the problem in general with anyone until the last person has returned.
3. Have each person leave the room one at a time to perform the measurement exercise.
4. When everyone has completed their measuring, write each of their results on the board. Chances are the answers will differ slightly. (You can encourage different answers by supplying a board which is slightly non-square.)
5. Discuss the process and the results with the group.

Here are some assumptions group members may have made as they measured the board. Consider including them in your discussion as they provide interesting and useful insight into measurement in general and are all potential sources of error.

Assumption: The board is symmetric.
Many people will assume this in their measuring, but few will qualify their result by stating this assumption. As I mentioned, you can encourage different answers by cutting the board so that it is slightly asymmetric.

The board should be long enough so that the difference is not readily apparent. The experience of your group in doing construction may play a part here also.

Lesson: A single measurement may not accurately describe the quantity you need to measure.

Assumption: The scale is accurate.

People generally assume the measuring tool, in this case the scale on the ruler or tape, is accurate. To bias the results, you could have given the participants a distorted ruler. If a measurement is worth recording, then it is worth recording accurately. To validate the accuracy of the ruler, a benchmark measurement should have been made.

Lesson: Validate your measurement tools.

Assumption: The measurement device is compatible with the measurement object.

(a.k.a. When all you have is a hammer, everything looks like a nail.)

If a six-inch ruler is supplied to measure a board six feet long, compatibility becomes a source of potential error. Errors from this assumption come from using one measurement device to measure a property for which it is not best suited.

Lesson: Identify and use the right tool for each measurement job.

Assumption: Length is the long dimension.

The instructions did not state which direction was defined as length. The person taking the reading made that selection as an on-the-spot decision. This error source is a potentially serious concern because the measurer believes his or her assumption is accurate and will base other parts of the process and conclusions on it. In industrial applications, this illusion can cause catastrophic consequences. The person requesting and receiving the measurement can believe that the value represents one fact, while the measurer believes with just as much conviction that the value means something else. This dichotomy can be disastrous.

Lesson: Clearly define measurement parameters to all who will be part of the process.

Assumption: The time required to perform the measurement is independent of the process.

In this experiment, if you only allowed a few seconds to read the tape measure the accuracy of the results would suffer, especially if the scale was not easily readable or the observer had little experience in using it. If you allowed more time, accuracy in reading the scale would most likely increase. At some point, however, the Law of Diminishing Returns will take effect and additional time will no longer result in increased accuracy. In fact, the accuracy of the reading can actually decrease if too much time is allowed and boredom sets in.

Lesson: Design each measurement process to challenge the measurer without overburdening him or her.

Assumption: The person read the scale correctly.

This assumption may appear trivial, but it can be the most important aspect of the measurement process. The observer is an integral part of the process. There is always a human at the end of any measurement process who must either record or interpret the observable parameter. Whether it is a simple ruler, a digital readout from an LED display, or a series of lights and sounds from a panel, the human must "read the scale correctly." Without accuracy in this part of the process, the technological precision of the measurement is useless. In many cases, the real variability in results does not come from any of the sources of error we've mentioned, but from the transfer of data by the human observer. There are many cases where technology has performed an accurate measurement and a failure of tragic proportions has occurred because the human observers either misread or misinterpreted the data. This is especially a hazard in operations when human fatigue or alertness plays a major role in system reliability [1].

Lesson: Design measurement processes with an understanding that the human being is the least predictable component.

Six Laws of Measurement

Now that we've looked at some of the potential pitfalls of measurement, let's examine some of its fundamental laws.

Law #1: Anything can be measured.

Technology, coupled with human creativity and the powers of mathematics and statistics, provides the potential to measure virtually anything and everything. Even the most seemingly subjective processes can be measured once the desired results are clearly defined. This means that the real measurement decisions are now much more difficult than in the past, and brings us immediately to our second law.

Law #2: Just because you can measure something doesn't mean you should.

With the ability to measure so much, it is increasingly important that we define exactly what should be measured and what we will do with the measurement results. Here lies the essence of what constitutes an effective measurement program. The difference between a successful measurement program and others is *not* what technology tools are applied and how much data are gathered. It is the manner in which tools are applied and the data are used. The person, team, or company that best understands their goals and the measurements necessary to steer toward them succeeds, while others empowered with the same tools may fail. Just as it is possible to eat all day, yet by eating "junk" foods be undernourished, it is possible to measure all day and receive little benefit. To carry the analogy further, just as junk food can be harmful, so can "junk measurements." We will see this further in our discussion.

My 5-year-old son is delighted every time he is allowed to use my measuring tape. He accepts this tool without question and proceeds to measure everything in sight. He has neither use for, nor understanding of any of his "measurements" but the device is fun to use. This may remind you of certain people when they are given new tools.

Consider a more practical example. If you want to measure availability of a particular system, and availability is computed from operating and repair times, there is neither reason nor need for additional measures such as mean time to repair or mean time between failures. The essence of these metrics is already encompassed in the desired metric, availability. Measuring additional variables related to availability requires more time and expense, provides no additional insight regarding the immediate need, and generally obscures the meaning of what you were trying to measure in the first place.

Law #3: Every measurement process contains error.

It is important to remember that the act of obtaining a measurement is a process *in itself.* Even though measurements may be accurate, the overall results of the measurement process may be wrong. There are always some errors involved in the measurement process. Success in measuring requires an understanding of the errors and the magnitude of each. Occasionally, error values turn out to be larger than the value of the measurement! In this case, a reevaluation (and perhaps a redesign) of the measurement process is needed. In any case, for both engineering and business, accurate error computations can be more important than the actual value of the quantity measured. Remember our experiment—in measurement, errors are very easy to make and generally difficult to spot.

Law #4: Every measurement carries the potential for changing the system.

Taking a measurement in a system without altering it is like getting something for nothing; it just plain doesn't happen. In physics, this law is called the Heisenberg Uncertainty Principle [2]. It says that if we try to measure the position of an electron exactly, the interaction disturbs the electron so much that we cannot measure anything about its momentum. Conversely, if we exactly measure an electron's momentum, then we cannot measure its position. Clearly measurement affects the process here. Even when dealing with the fundamental building blocks of matter, measurement tampers with performance.

It is virtually impossible to have any measuring process that does not affect the process being measured. For a more common example, consider a pressure valve in a segment of piping or an automobile tire. To move the needle to the calibrated distance corresponding to the internal pressure, kinetic energy must be released, which in turn reduces the pressure to be measured. Although the needle mechanism reduces the pressure a negligible amount, it nevertheless is reduced.

Here is another example. Nuclear power plants rely on large diesel-powered generators to supply emergency electrical power for reactor operation. The Nuclear Regulatory Commission requires that these generators be started once a month to ensure that they are in good working order. In reality, the testing itself causes wear of many of the components. It is likely that testing at this high frequency actually causes the diesel engines to be more unreliable. Some people say that the diesels should be started and kept running. The controversy in this area goes on, but the paradox is real—excessive testing to give the user confidence that the equipment is reliable can actually make the equipment more unreliable [3].

In business and engineering applications, this law also recognizes how people react to measurement. Consider the difference in telephone transaction styles between the operators who provide telephone directory assistance and those who answer calls to exclusive catalogue order services. It's easy to tell which of the two groups is measured on number of calls they handle in a day. Each of the groups is providing high *quality* service by delivering exactly what the customer expects, yet behaviors differ based on the measurements selected to achieve desired business results. Imagine the changes that would occur if performance of Neiman Marcus operators was measured solely by the number of calls per hour each handled!

I recently visited a large corporation where a very rigorous measurement plan is used to gather performance information for each of their plants. Inspectors are sent to check many components of processing and compile a plant-specific "quality score." Hoping to observe "typical" plant conditions, inspections are made without advance warning. However, knowing that the inspectors visit each plant about once every 18 months, the company's employees prepare for measurement. Maintenance logs show a flurry of activity during the weeks immediately preceding inspections. Although this probably raises quality scores for the plants, it is certainly not an effective way to schedule maintenance!

This law is often overlooked. Keep in mind, especially when the measuring process involves human responses, the results are biased to some degree by the measurement process itself. Exact measurement should be a noble goal in every endeavor, but keep in mind it is an ideal. Design of the measurement parameters and process can take advantage of this by prioritizing measurement of desired actions or critical components. Here is a situation where "junk measurements" can distract from the task at hand by focusing attention on more trivial situations.

Law #5: The human is an integral part of the measurement *process.*

There is absolutely no doubt that humans affect and influence a measurement process. However, the specific ways in which it occurs are not always clear. Some pertinent topics of this subject are discussed in more detail in Chapters 10 and 11. For now, I want only to discuss two specific aspects that are connected directly to measurement programs. The first topic addresses how people receive and process data into information. The second subject identifies how employees respond to measurement.

Like beauty, the value of information is in the eyes of the beholder. Collecting data and transforming the facts and figures into information require human intervention, judgment, and control. The observer can enhance or degrade the timely, accurate, and complete recording and compilation of data, depending on factors such as personal interests, motivations, level of competence, distractions, and fatigue. The degree of difficulty in identifying information and extracting it from the data will vary depending on the complexity of the system you are measuring. A relatively simple device such as a smoke detector is designed to transfer airborne particulate data into information in the form of a piercing alarm. If you have ever had a smoke alarm go off unexpectedly, you will agree that it is very unlikely that anyone would misinterpret the information being transmitted from the device. Unfortunately, there are many situations where the data information processing is not as straightforward. Even with the standardization of data entry, calculation, and information presentation by computer systems, the ultimate *interpretations* of information and subsequent actions are largely left to the people in charge. The human is generally the final and most crucial link in the chain, and is the source of that large group of unplanned events attributed to "human error." As computer software becomes more sophisticated, it can provide responses that are more standardized and predictable. In the foreseeable future, though, the human element of the data information process has a long way to go.

There is another subtle, but real, effect that humans can have on a measurement process. To explain what this is and the powerful effects that have been observed, we'll go back in time. Between 1927 and 1932, a productivity study was performed at the Hawthorne plant of the Western Electric Company outside Chicago, Illinois, by Professor Elton Mayo of the Harvard Business School. Mayo's plan was to study the effects and test the predicted correlation between overhead lighting and employee output. The women working in the Relay Assembly Test Room made management history, changing forever how supervisors and company leaders view workers' motivations, group dynamics, and autonomy. The study led to a quite unexpected result. Professor Mayo found that worker productivity kept rising *regardless* of the light intensity. *The study concluded that people work harder when you kept introducing change into their environment.* Since this landmark study, the observed behavior is called the Hawthorne Effect [4-6].

When employees are cognizant of the measurement process or even just management's intent to measure, improvement is often seen before any substantial changes are made. Employee knowledge or awareness of the measurement process affecting them either directly or indirectly *introduces change into their environment.* This will produce a wave of improvement that subsides if not supported by substantive measurement and feedback. After the newness wears off, things will go back to "normal" without additional changes to the process being measured. I have seen the Hawthorne Effect in action. Companies have made sizable reductions in costs and increased their productivity without any major changes in technologies or changes in their business processes. The improvements occurred simply because employees became aware that a measurement process was beginning.

Once you are aware of the Hawthorne Effect, you can use it to your advantage and receive benefits more quickly than expected. Just remember, the initial improvement is like riding a wave. After a short time it will fade away or level off unless the measurement program re-introduces substantive changes into the business or industrial management process.

Summary Law #6 (Jones' Law): You are what you measure.

Every structure is defined by its individual boards or beams and their relationships to each other. Structural forces are distributed among the primary and secondary beams to offer an integrated structure with forces uniformly distributed throughout. Any concentration of forces on a single beam indicates an improper design. This is an appropriate analogy to consider in developing a measurement strategy. You can consider each single measurement quantity a beam. By itself, each has limited meaning or definition. Together however, they can define a simple shack, a cathedral, or your business direction, depending on the needs and resources of the architect and builder. This perspective must be taken to ensure your measurement strategy is concise and effective. Your set of measurement variables should be complementary, reflecting exactly what you want to know from the measurement and your unique needs. Here is the universal guideline that should be used in all measurement development:

> **Each measurement must be directly related to achieving the mission of the organization.**

Any measurement that doesn't fit this criteria is a "junk measurement—satisfying for the moment, but in the long run distracting and potentially harmful.

The process of measuring people acts as a stimulus; after being subjected to it in a fairly periodic manner, it elicits an auto-response. The more you measure, the more pronounced the response. Measurement can and does affect the process. In some cases, this is precisely what is desired as the measurement system can and should be used to elicit positive behavior. Sometimes it is highly desirable to have the measurement change the system. Make sure that you consider the changes resulting from the measurement process as you plan your measurement strategy.

Two Additional Issues: Measurement Bias and Measurement Imbalance

Designing measurements in a manner analogous to the primary and supporting beams of a structure was suggested for more than intuitive reasons. It will also help avoid biases and imbalances that may surface in the results.

First, let's address the issue of *measurement bias*. The entire measurement process provides an overall view resulting from many individual characteristics. Primary and secondary metrics must be constructed to help safeguard from the error associated with drawing incorrect conclusions from individual measurements. You've often heard the expression "the whole is greater than the sum of the parts." This certainly holds true for process measurement. Be careful not to unintentionally bias your results by emphasizing individual measures rather than the overall result. Again, to use the building analogy, remember the structure, not just the beam.

Measurement imbalance is produced when one metric or set of metrics supplies information that is either too detailed or too profuse, overwhelming the level or amount of information from other measurement parameters. Information from this area may begin to dominate the measurement results. For example, consider a control panel where all readouts are important, yet one is a large LED showing information detailed to three decimal places, while the other readings are displayed in standard analog form. Simply by virtue of visibility and apparent precision, the former indicator would dominate the measures. Information imbalance obscures the overall measurement process and is often indicative of either a concern over a particular area or simply more readily available information from a particular process. It should not be an unintended part of the measurement process.

Developing a Measurement Strategy

Our universal measurement guideline defined the need to directly align a measurement strategy with the mission of the organization. This requirement cannot be over-emphasized. Today, and even more so in the future, the ability of a company to manage its measurement processes is essential. Progress toward continuous improvements in quality, productivity, delivery of services, and profitability has little meaning without an effective measurement strategy in place. Data collection, storage, retrieval, and processing are fundamental to today's businesses. It's easy to collect data. A measurement strategy is the plan for gathering the data you need and distilling it to produce the type and quality of information that will help determine the directions of the business. Measurement without such a strategy is a costly bit of sound and fury signifying little or nothing.

Measurement is and always has been an art, and little has changed with the application of technology. In many ways technology magnifies the power of this statement. It provides the user with more choices for measurement, more types of measurement devices, more data from which to extract information, more capability for precision, more opportunity to succeed and more opportunity to fail.

In short, each enterprise must understand its bottom line needs and measure only the factors that reveal whether or not anticipated results will be achieved as expected. The right measurement strategy allows you to best monitor and modify activities in your business so that "random" failures do not occur.

References

1. Moore-Ede, M., *The Twenty-Four Hour Society,* Addison-Wesley, New York, 1992.
2. Gerjuoy, E., "Uncertainty Principle," *McGraw-Hill Encyclopedia of Science & Technology,* 7th Edition, 1992, Vol. 19, p. 22.
3. Jones, R., B. and Bickel, J., H., "Optimal Test Intervals of Standby Components Based on Actual Plant-Specific Data," Probabilistic Safety Assessment and Risk Management, PSA '87, Zurich, 1987, Vol. I, pp. 209–212.
4. Neimark, J., "Shake it: The Hawthorne Effect Established the Overwhelming Importance of Newness," *SUCCESS,* July/August, 1988, pp. 48–49.
5. Geber, B., "The Hawthorne Effect: Orwell or Buscaglia?" *TRAINING,* Vol. 23 November, 1986, pp. 113–114.
6. Parsons, H. M., "Hawthorne: An Early OBM Experiment," *Journal of Organizational Behavior Management,* Vol. 12 1992, pp. 27–43.

CHAPTER 3

Responsible Statistics

Statistics: Numbers looking for an argument.

Changing Times Magazine

Several years ago, I was giving a briefing in Washington to a group representing a regulatory branch of the government. They were reviewing work on determining public evacuation strategies from pollutants released as a result of industrial accidents. During my presentation, I showed a graph depicting the probability of various levels of consequences from the dispersion, transport, and environmental damage. This curve was similar to the one shown in Figure 3-1.

As the graph shows, there was a very small probability of severe accidents. However, there was one possible accident sequence where the pollutants went into the atmosphere, were transported by the prevailing winds over a dense population area, and then deposited on a large city by rain or snow. This situation produced a point on the graph far away from the other results. All of the data points were connected by a straight line

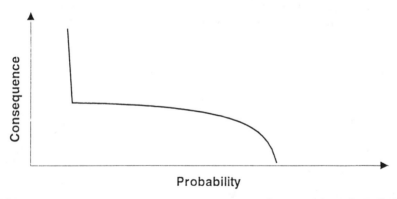

Figure 3-1. Probability vs. consequence for accidental pollution release.

26

so there was no way for the audience to know that the large shift in the curve was caused by only one point. One member of the audience asked me to explain the reason for this seemingly strange behavior on the graph. I proceeded to explain the particular sequence of events that produced this result. To help the questioner understand the meaning of this statistical outlier, I equated this event with the probability of having a large meteor fall on a densely populated area. As a matter of course, I further explained the high degree of error in this calculation and that the graph was more for a general perspective than specific policy planning.

The questioner understood the result more than I expected, as shown by his second question, one that I will never forget. The question was:

"I understand from your explanation, Dr. Jones, that the scenario is very, very unlikely, but can it happen tomorrow?

"Yes," I responded.

Proper Use of Statistics

Statistics do not prove anything. The field of statistics is incapable of this because the process of proving something requires the exact identification of strict rules that work *all* of the time. Unlike statistics, mathematics *can* be used to prove relationships. Its logical construction forms a perfect framework of theorems, corollaries, and lemmas designed for the purpose of absolute, perfect proof. Mathematical conclusions are ideal, complete, and absolute. Given a certain set of conditions, either a relationship is true or it is not. There are no other choices. Statistics, on the other hand, help people make decisions in the non-ideal world we call reality. It deals with imperfect data, and supplies information that assists in, but does not completely perform, decision processes.

The field of statistics is very broad. We will restrict ourselves to the segment called inference, or trying to deduce information from data. Basically, statistical inference is a method of using data to *support* or *not support* hypotheses. A hypothesis is a statement postulating some relationship in a process. Hypotheses can be very simple or very esoteric depending upon the phenomena under study. Here are some typical examples of hypotheses:

Hypothesis: People who eat a low cholesterol diet have fewer heart attacks than people who eat a high cholesterol diet.

Hypothesis: People who exercise regularly have fewer heart attacks than people who do not.

Hypothesis: People who do not smoke live longer than people who do smoke.

Hypothesis: People who drink red wine with meals are less likely to have heart attacks than people who do not.

These are all statistically true statements. However, if taken as potential mathematical theorems, they would all be proved to be false. Why? For a mathematical proof to exist, it has to be true 100% of the time. In each of the hypotheses listed, a mathematician could prove each statement false by finding just one exception. Even though these statements are not mathematical theorems, they are generally true since as a group, most people who conform to the guidelines will experience the stated results. It is this area of bringing quantitative information to practical applications where statistics plays a major role. The only drawback is that the results are not always correct. If you are concerned about the behavior of groups and not about the behavior of individual items, then statistics offers many powerful tools. If your concern is oriented towards the behavior of individuals, then other modeling or simulation [1] methods are applicable. Remember, statistics describe the *rule;* mathematics the *absolute.*

We exist in a world where we are exposed daily to a large amount of information in our professional and personal lives. Much of it is accompanied by "the numbers" that support the presenter's point of view. Together with charts and graphs, these "numbers" can provide apparently compelling evidence to support or reject the hypothesis at hand within stated limits. There is a general perception that the more quantitative or "scientific" the measurement tools, the more likely the result is correct. When people use numbers in a convincing manner, it is easy to suffer the illusion of accuracy. The mystique of mathematics and statistics lends itself very well to abuse as the general population is poorly equipped to argue or even question "the numbers."

When it comes to the proper use of statistics, there are serious pitfalls that many people, including the press, encounter. The use of statistically based metrics and conclusions based upon the application of statistical tools represents a powerful and convincing medium. These benefits are, however, accompanied by a responsibility to communicate effectively, clearly, and completely.

Consider some examples that illustrate the common misuses or abuses of statistics. As you read them, focus on some of the ideas they present for measuring the quality or worth of statistical arguments and conclusions that you might see or hear.

Example #1: The Doctor's Consultation

Suppose you go to a medical doctor for diagnosis and treatment of symptoms and, after a series of tests you meet with the doctor to discuss the results. The doctor says there is good news and bad news. On the bad side, he/she regrets to inform you that the tests have revealed a serious disease. On the good side, the doctor informs you that the likelihood of a cure is 70%. You walk out of the office shocked, but feel encouraged about your chances that you will live. After a long treatment period, you die.

In retrospect, your personal "probability of cure" was zero. What happened? The doctor was correct in stating that the probability of a cure was 70%. This means that 70% of some group of people that some doctors have diagnosed over some time interval in some area of the country or world with the disease survive. You, on the other hand, are one person with one outcome. Individual behavior is not predicted from statistical arguments.

This type of statistical misinterpretation or miscommunication is commonplace. It represents a major concern because, using the example, the doctor is telling you "correct" facts, but they represent incomplete information from a personal, individual perspective. People base their actions on an "illusion of accuracy" when, in fact, nothing could be further from the truth. This type of misuse of statistics certainly is not limited to the medical profession. It is even more common throughout the rest of our society, including in industrial conference rooms.

Example #2: Health Risks of Cellular Phones

Not long ago there was a national media outpouring of articles warning that electromagnetic radiation given off by portable cellular phones could cause brain cancer. The claim was not based upon scientific research, but on two lawsuits filed by the families of two people who used these devices and died from brain cancer. If you apply the normal incidence rate of brain cancer to the millions of people who use these devices you would expect on the order of several hundred people to develop this type of cancer anyway. Is there a connection between cellular phones and brain cancer? The EPA is investigating, but at this point, there is no way to answer this question with an absolute yes or no. What can be said is that there is no conclusive evidence that cellular phones are dangerous.

Example #3: Cancer Risk of Vasectomies

This example is based upon scientific evidence and illustrates how conclusions can be flawed by not considering the effects of factors omitted from the analysis. Two medical studies found that men who have had vasectomies had almost twice the normal incidence rate of prostate cancer. Should you conclude that there is some connection between vasectomies and prostate cancer? Before you decide, read on. The research effort looked at men who had the procedure performed more than 20 years ago when it was relatively new. If these men were willing to submit to a new surgical procedure, chances are they were also more conscientious in receiving medical checkups where prostate cancer can be detected. Furthermore, today's medical research has not identified any mechanisms that can connect the vasectomy effects to prostate cancer. The point is that it may be inaccurate to apply statistically sound results to the general population because the study's sample may not represent the general population.

Example #4: Marriage & Divorce

"Fifty percent of all marriages end with divorce." Right? This commonly quoted statistic also has population factors at the root of its misuse. First of all, the quote is not true. It stems from the fact that the number of divorces is about 50% of the number of marriages per year. In other words, one half of all couples that take wedding vows get divorced. No. To see the flaw in this reasoning, just examine how many people are available to get married compared to the number of people (or couples) that are available to get divorced. More than 75% of the adult population is married. It is not surprising that the frequency of divorces from this large pool of people is greater than the frequency of marriages from the relatively small pool of available single people. If you consider the differences in population sizes, it turns out that almost 90% of all marriages survive [2].

Example #5: Handguns Are Dangerous

When people try to prove or disprove a perceived fact with numbers, the stage is set for both sides to exchange numerical blows and the whole process can often be characterized by Shakespeare's Macbeth's quote, ". . . full of sound and fury signifying nothing." Here's one such example. A proponent of gun control claims that 24,000 Americans are killed each year with handguns. This statement is true. However, opponents of gun control respond by also accurately stating that almost 55% of all gun deaths are attributed to suicide and fewer than 0.2% of existing firearms are used to commit crimes [3]. Are you convinced of anything?

Statistical abuse and misuse is common in the public media. It has become so frequent that the abuse has been given a name: Statspeak [4]. This name is derived from the similar usage of the English language, doublespeak. Literally every newspaper contains examples of either errors by omission of facts or errors in statistics application. In business, one can only speculate the extent to which it exists. The main reason for the rapid increase in Statspeak is the proliferation of statistical analysis capabilities readily available in spreadsheet and other PC-based software. People now have the ability to use statistical analysis tools with little or no formal training in statistics. It is very easy to use these powerful tools for the wrong job. Unfortunately, this results in convincing-looking arguments that may be based on gross error. Too many conclusions are made from a statistical analysis in which the underlying assumptions are neither verified nor stated. Such conclusions are tenuous at best.

Guidelines for the Correct Reporting of Statistics

The following set of guidelines outlines the components that should be clearly identified in the reporting of statistics or identified when using statistics to make a case for or against a certain hypothesis. While it is not possible to identify all of the variations or specific components for each application area, every statistic reported should contain elements from the following four essential areas [5].

1. Number: How many items were counted in forming the statistic?
2. Phenomenon: What type of item was counted?
3. Time Frame: Over what amount of time was data collected?
4. Geographical Limitations: What region, state, or general physical space was used for data collection?

Consider again the statistic used in Example 1 "The Doctor's Consultation" that we just discussed: "70% of people who are diagnosed with the disease are cured." Let's take a look at the guidelines for the correct reporting of statistics from the perspective of this example.

Number

How many people are in the group from which this claim is being stated? 10, 100, 100,000? The situation of only 10 people ever having the disease and 7 were cured is much different that 100,000 people having the disease and 70,000 being cured.

Phenomenon

Does this disease affect men and women the same? Is there a difference in cure rate with age, lifestyle, body weight, blood type, or work environment? Were all of these varieties of people included in the group?

Time Frame

Has the data base from which the 70% cure rate been accumulated over 50 years, 10 years, 1 year, or 6 months?

Geographical Limitations

Was the data selected on a world, national, or regional basis? Is there a difference in cure rate with living location?

If you apply these guidelines to statistical claims you see in commercial advertising or read in the newspaper, you will be surprised as to the extent of Statspeak in our society. Take a look at how little you really know about the basis for these claims which are common in advertising [6]:

- "XYZ is preferred 2 to 1 over the other brand . . ."
- 3 out of 4 doctors recommend . . ."
- "9 out of 10 people surveyed said . . ."

The important point here is that statistics has an important role to play in transforming data into information. However, care must be exercised to ensure an accurate and complete understanding of statistical validity and limitations.

Difficulties in Using Failure Data

Statistics is a field that uses the logic of mathematics to transform incomplete, imperfect data into usable information. Equipment and systems in an industrial setting, regardless of the business, are subjected to factors not accounted for in the rigors of mathematical modeling. The term *synergy* is appropriate here because the factors taken by themselves may not be as important as their combined effect on system time-dependent reliability. There are always synergistic relationships that influence availability and failure frequency, regardless of the process. Do statistical models of reliability have application? Yes, but before these powerful techniques can be applied, the potential for the influence of synergistic, unrecognized factors of equipment operation must be investigated.

What is "real world synergy"? The following are some examples of real-world synergy affecting equipment reliability. This list is in no special order.

- The failure causes degradation of other system components.
- Repair work does not fully correct problem.
- The parts used in the repair are not of the same quality as the original parts.
- Some failures require part replacements and others require only adjustment.
- Replacement parts may be damaged or have deteriorated in shipment.
- Replacement parts may be damaged or have deteriorated in inventory storage.
- The failure caused additional failures of other system elements.
- The failure was due to accepted operation-related practices.
- The failure was due to operation error.
- The failure was due to human-machine influences (e.g., spilling a can of soda that was left on a system control by inspection personnel).
- Overhaul does not restore the system to original condition.
- System demand changes in time as product demand fluctuates.
- Physical system environment or product corrosiveness differs although systems are similar or identical.
- Engineering system design changes.
- Time-based preventive maintenance is performed irregularly.
- There were condition-based predictive maintenance data collection errors.
- Predictive maintenance is performed irregularly.
- Changes in maintenance management resulted in a staff with more or less system experience.
- Changes occurred in data recording practices (e.g., implementing a computerized maintenance management system).
- Too much maintenance.
- Too little maintenance.

This list itself could be lengthened into a book. The point is that reliability modeling and prediction using statistical analysis of available data contains many assumptions.

A natural question at this point is "How can synergistic factors that affect equipment availability be investigated if we do not know what they are?" The answer: Tests can and should be performed to determine if real-world synergy effects can be measured in data. Let's look at some possible ways to detect synergy.

Sneak Analysis

The effects of real-world synergy are often very complex and difficult to quantify. NASA has developed a special methodology to study reliability and safety in these cases called Sneak Analysis [7,8].

Sneak Analysis originated in 1967 for the manned space program after the Apollo I capsule fire. NASA became increasingly concerned with their ability to adequately ensure the safety and reliability of systems. In addition, they were experiencing a growing number of unintended, unexplained, untimely events.

A "sneak condition" is a latent hardware, software, or integrated condition that can cause unwanted events to occur in the system. Symptoms of sneak conditions are problems that are non-repetitive in tests or simulations, problems that more conventional analyses cannot detect and produce higher unexplained failure rates than expected. This situation is termed a "bug" when it occurs in computer hardware or software programs. In general, sneak conditions occur when programs and processes go awry intermittently without apparent cause.

There are three major factors that cause sneak conditions:

1. *Spaghetti Factor:* This factor deals with the entanglement of electronic, software, and functional paths. Size and complexity contribute largely to this type of sneak condition.

2. *Tunnel Vision Factor:* System design divided the process into discrete areas. This division segregates the overall design and development so that complete, detailed operation requires many interfaces.

3. *Frame of Reference Factor (a.k.a. A Rose is* not *a Rose):* This aspect describes how different people can interpret the same information differently. For example, two operators can interpret imprecise (or sometimes even precise) instructions in different ways. Also, communication between design departments may seem satisfactory on the surface, but found inadequate only when the subsystem interface design is wrong.

Sneak Analysis has applications with great potential in the general reliability analyses of complex systems. The field is very young and has primarily been applied to the space program so far. Information is available from NASA to enable its application in other fields. Current industrial efforts are in the research and development stage.

Prediction Modeling Approaches

The preceding discussions have identified how many factors can influence system reliability in actual practice. The reliability of systems in the field is not simply related to reliability measurement in the controlled environment of a laboratory. In addition, the net behavior of a group of components is neither simple, nor intuitive, nor easily predictable from standard reliability models. One conclusion that can be made from the list of "Real-World Synergy Affecting Equipment Reliability" cited earlier is that some factors influencing failure frequency and consequence cannot be measured from a practical perspective. How then can meaningful reliability estimates be determined with all of these complicating factors? These "complicating factors" are no more than real-world influences that have existed and probably will always exist. The point of this discussion is that textbook solutions to real-world problems may or may not be applicable to any given situation. If the assumptions used in the techniques cannot be verified, then it is hard to place any confidence in the results. This lesson may sound elementary, yet its violation is common.

As we mentioned earlier, a significant part of the problem is the innocent, but nevertheless inappropriate use of statistical computer software. Sometimes people become so relieved at finding a tool that seems to fit their needs or become so enamored with a familiar or flashy tool that they can overlook the basics. In my experience, this is especially true in the application of statistical methods. The process of validating the assumptions (and hence the validity) of both statistical and mathematical methods is often a very difficult problem by itself. It can be more difficult to verify the assumptions of a model than to do the modeling itself.

Let's restrict our discussion to the reliability of equipment and systems. In these cases there is typically a relatively small data base because there has been a small number of failures. The reason for this is a matter of practicality and economics. If a system has a large number of failures that are affecting production or causing injuries, chances are the system is being fixed or redesigned. If a system has a large number of failures and this behavior is tolerated, chances are that its contribution to production is relatively small and a reliability study would not be economically justifiable anyway. This brings us back to the case of the small number of failures. Without sufficient data it is often practically difficult to make any specific modeling verification. You just have to use your best judgment and keep this in mind when reviewing the results.

Even with a small number of failures, when it comes to forecasting failure events, we *are* capable of predicting the future. Making decisions in industry is somewhat similar to playing games of chance. In both situations, you can have a pretty good idea of the odds for and against each choice. However, in industry there is usually only *one* opportunity to get it right. From the information that is available, people (not computers) must decide issues that can produce financial success or serious financial loss. Being right "most of the time" is not enough. The good news is that, in industry, if your data is accurate and your methods are right, then you've significantly stacked the deck in your favor.

Unfortunately, there is no mathematical formula for 100% success. Numbers, figures, and computers play a growing role in assisting people in their decision processes. As with measurement, it is not *what* tools you apply to assist in decision processes—it is *how* you apply them. For the most part, everyone has the same tools anyway.

There are two basic types of predictions that come out of mathematical modeling: Deterministic and Statistical. To illustrate these approaches, consider the task of predicting when a light bulb will fail.

The *deterministic approach* to this problem is to take one light bulb and perform extensive analyses of its components and develop a mathematical description of the processes that are operating when the bulb is lit and when it is off. This method uses mathematics to describe the detailed physical changes of the light bulb's constituent parts and, from this description, determines the time when the bulb will fail. This is called reliability modeling. It models the time to failure "from the inside." As you can conclude from this short discussion, detailed knowledge of a system's internal processes is required for using this method.

The reality is that in most situations, exact information of the detailed processes is not available, so reliability modeling can't be done. The operating environments of industrial equipment generally are not the same as laboratory test conditions. Consequently, there is no real knowledge of the specific detailed processes affecting each piece of equipment. Trying to deterministically model system function and degradation to failure is neither possible nor practical.

The *statistical approach* is best described as "modeling from the outside." Here's how we would use it for our light bulb example. Gather a large number of new identical light bulbs and place them in one room. Turn all of them on at once and keep track of when each light bulb fails.

From these records, a histogram can be made showing the number of light bulbs failing between each given time interval, such as the number failing between 100 and 110 hours, the number failing between 110 and 120 hours, and so on. By dividing each histogram column by the total number of light bulbs, a probability distribution can be formed. It will show the probability of a light bulb failing between different time intervals of operation. From this distribution, the mean time to failure and a host of other statistical quantities can be calculated. Each one of these results gives the observer a piece of information.

Because no two light bulbs are exactly alike, they fail at different times. There is no exact time when all bulbs will fail. The distribution of failure times allows the person involved in determining light bulb failure time to choose the most representative time for their application. Historically, the mean, median, or mode values have been used. Other descriptive statistics are also available to help describe the process. Because there is no exact answer to when all light bulbs fail, probability and statistics provide ways of supplying information to assist in deciding on the best indicator of light bulb lifetime.

Recognition of the realities in the industrial environment where policy decisions must be made is absolutely imperative. If being right "most of the time" was acceptable in policy decision-making, then computers could be programmed to compute various policy alternative solutions and choose the best one. When only one decision opportunity is present and being right most of the time is not acceptable, probability and statistics can provide decision makers with valuable information to *assist* in the decision process. The decision still must be made by people.

Remember, probability theory began with gambling problems in the seventeenth century. Statistics also began at this time with the compilation of mortality tables. Probability and statistics are not designed to predict exact behavior, but rather the overall behavior pattern of an ensemble or population. When you use these tools, recognize that their results are not exact. There is always a chance that policy results or observed failure times will be different from the prediction. When you use probability and statistics you *are* gambling. Make sure you know the odds.

Probability Modeling Assumptions: Independent & Identically Distributed

Probability and statistics are powerful tools in the hands of the right people. Like any other tool, however, they have limitations and violating these constraints leads to errors. The major difficulty is that the user of these tools may not know if any underlying statistical assumptions were violated and may subsequently apply calculations incorrectly. This is another example of the saying "Garbage In, Garbage Out." The trouble is the user may not know he/she is dealing with garbage.

In dealing with numbers, it is usually not obvious whether or not the mathematical tools are appropriate or inappropriate for the numerical data being analyzed. If a physical tool is used for an application for which it wasn't designed, the user knows immediately and adjustments can be made to the solution process. For example, if a mechanic needs to tighten a nut to a specific torque, there is only one tool that will fit the task: a torque wrench and a socket of specific size. The tool matches the task requirements. If an adjustable wrench is used instead, it is obvious that task requirements and the applied tools are not suitable. When dealing with numbers, the suitability of the mathematical tools is not as easily determined. But before you can choose the right tool, you must be familiar with what you're working on—in our case, we're working on data.

Industrial data can be generally divided into three categories:

1. Operation or process data: required for system control
2. Equipment condition data: obtained from predictive maintenance (PdM) analyses
3. Failure data

Operational and process data are measures of equipment function to control its contribution to the production process. Equipment condition data are a measure of various performance characteristics that together determine equipment failure potential. Failure data document when things did not work.

All of these measurements are important. The first two are measurements of success—everything is working. The last measures failure. Measuring failure frequency and related effects is essential to continuously improving availability and safety, and risk reduction. It is this area of analyzing failure data that will be addressed next.

Occurrences of failures will follow one of two patterns: either failures will be related to one another, or failures will be completely unrelated and independent. A major area for improvement in industrial reliability is to NOT assume that one or the other of these properties is satisfied without examination of the data. Without considering these factors before applying standard probabilistic and statistical methods, the validity of any results is questionable.

In the real world, very little that happens to systems, including their failures, is independent. For example, equipment and systems are aging, failure mechanisms are changing, condition monitoring or PdM may intervene before many failures occur. Even the concept of what actually constitutes "failure" can change. And let's not forget that people continuously interact with the equipment in both positive and negative ways. Listed earlier were some causality relationships that can exist; there are many more.

There are situations where failures are completely unrelated. This is usually because insufficient data are present to totally confirm or reject a relationship between the failures. Even if sufficient data are compiled, causal factors can cancel each other. We will first discuss the case where failures are independent, and a related second assumption, that they are identically distributed. These two basic assumptions regard failure data that must either be presumed without analysis or determined by testing before standard probability methods can be applied. They state that the data are *independent* from each other and are from *identically distributed* probabilistic processes. These assumptions, commonly abbreviated as IID, form the foundation for the application of the usual statistical analysis tools.

Independent Data

Data are independent if there is no association between them. For this discussion we will specifically deal with the times between failures of equipment and systems. An illustrative and interesting example of dependent data occurs when a relatively long time span between failure data values is generally followed by a short time span between failures. This behavior is characteristic of maintenance-induced failures. The system initially failed for some unknown reason. The problem was thought to be corrected, and the system was placed back in service. A second failure was then identified in the same or related component. Often, these types of failures are described as "adjustments" in describing the type of maintenance action applied. From a system perspective, the second failure is just as much a failure event as the first, yet the second is due to the human-machine interface, not to engineering or reliability degradation.

For another example, consider the operation of a motor-pump unit. The pump initially failed due to excessive leaking of a seal and was repaired the same day. The next week, another seal failed. Seal failures continued to plague this unit. About a month later, the motor bearings needed to be replaced. When the bearings were replaced, the alignment of the motor, shaft, and coupling were checked and found to be beyond specifications. Finally, the unit underwent realignment and was placed back into operation. The frequency of seal failures dropped to nearly zero.

From a mechanical perspective, the misalignment was probably why the seals were failing in the first place. If the shaft alignment had been checked when the first seal failure was observed, most likely the motor bearings would not have failed. These failures are *dependent* because there was a causal relationship among the failure mechanisms. This direct observation of reality is often faster, easier, and more accurate than the standard statistical tests for data independence.

Statistical tests treat data like some people perceive the government treats them, like a group of numbers. Statistical methods are not always the best way to ascertain if data are independent. To maintenance, operations, and management, failure and operating data are real. It is the history of their operating successes and failures over a given time period. It is possible that a peer review of failure data can ascertain whether or not causal factors among failure data can be identified. This type of testing for independence is not discussed in textbooks because of its non-analytical nature. Nevertheless, it can be of significant value in industrial environments. Dependence or independence among failure data can be best judged by people knowledgeable of the system operation rather than strictly with numerical tests that examine only the clinical behavior among the data values.

This is not to say that statistical tests are without merit. There is a significant advantage to using the statistical tests if the data set is suitable, i.e., large enough. Statistical tests do not have careers, axes to grind, good and bad days, prejudices, or any other traits that are in humans. Their clinical results can be of great value. Statistical tests always provide a probability level that gives the observer a degree of confidence in the results. They are never wrong. They only state that a certain relationship is true or not true at a given level of significance. The person accepting or rejecting this information *can,* however, be wrong. The relationships shown through the use of statistical tests can help determine areas of concern and, if nothing else, can be a basis for some relevant discussions among the people involved.

Independence among time-between-failure (TBF) values generally means that large TBFs are not necessarily followed by larger or smaller TBFs and vice versa. In other words, the values of future TBFs are not influenced by previous values. The independence assumption would clearly be violated if an age vs. reliability relationship existed. For example, in the case of perfect deterioration, every TBF would be less than the previous value. This dependence would signify the assumption of independence is not valid.

It is very common in reliability modeling efforts to omit testing for independence. One reason is the relatively large number of values required for adequate precision. More than 50, and preferably more than 100 values are desired. Also, these tests are relatively difficult when compared with trend testing for the identical distributed property. The main reason they are omitted, however, appears to be a lack of awareness that testing for independence is important. This leads us to the discussion of the second assumption.

Identically Distributed Data

Identically distributed means the probability distribution from which the TBFs are derived is not changing. For failure data where time, or some other related variable, such as cycles, is used, this means the same probability distribution form can be used to model the failure frequency for the time period under consideration. The assumption of identical distribution signifies that the chronological order of data values does not contain any information. There is no observable age vs. reliability relationship, or in other words, the system is not getting any better or any worse. We will discuss a simple trend analysis procedure later in this chapter to test for this property. Because there are many potential incentives for studying reliability estimation, we'll examine our specific interests in this area in the next section.

The Purpose of Reliability Estimation for Risk-Based Management

The general procedure to assess the validity of the IID properties for a data set is:

1. Perform a trend analysis examination to test failure data for the identical distributed statistical property.

2. If no trends are found, test the data values for independence using either statistics or peer review or both.

3. If some or all of the data values do not pass the test for independence, an investigation should be made to sort out the dependent values and begin the data analysis again.

4. If no dependence is observed, the data is IID, and the standard stationary methods of statistical reliability (such as Weibull or exponential) analysis can be applied.

The question addressed in this section is: "What information is desired from the data?" The reliability objective of this text is to develop a representative estimate of time between failures or, stating the goal another way, to develop a realistic measure for the probability of failure. If data are found to be time dependent (a trend), the representative time between failures is computed using the trend relationship observed in the data. Methods to compute trend estimates of this value are contained in the next sections. If stationary methods are applicable and the data set is sufficiently large, the data can be fit to the best failure probability distribution, such as exponential or Weibull.

With a distributional (IID) approach, other representative times can be estimated, for example, the *distribution mean* or t_{50}, which is the time for which the probability of failure at a time less than this is 50%. This statistic is also called the *median*. Another representative time that can be estimated with the IID approach is t_{95}, which is a conservative estimate for a representative time between failures. It is the time for which the probability of having failures less than this time is 0.95. Saying this another way, 95% of the failure times are less than this value.

With the distribution form defined by the IID data, there are many versions of the representative time between failures that can be easily computed. We use the Mean Time Between Failures (MTBF) as a representative time between failures when no trend is identified in the independent data. The next section discusses a procedure to detect trends in data sets and, in these cases, how to compute a representative time to next failure.

A Trend Analysis Procedure

In this section, the dimension of time is considered through the trend analysis of failure. A trend is a relationship between data values. For example, consider the following systems and their associated TBF data:

Time Between Failures (days)						
System A: 175	150	100	75	50	25	5
System B: 5	25	50	75	100	150	175

Both systems have the same MTBF and both have the same standard deviation. However, they are very different. For System A, the time between successive failures is getting smaller. System A's reliability is clearly deteriorating. A look at System B shows the time between failures is getting larger. Reliability for System B is improving. The time or chronological ordering of the failure data contains valuable information.

In actual practice, are trends rare? In the world of continuous improvement, where considerable effort is expended on systems to enhance performance and reduce failures, the answer is no. In fact, continuous improvement implies that improvement trends *should* exist.

The type of trends discussed here are statistical in nature. They may or may not be obvious to your intuition. One fact is certain: they should not be ignored by skipping the analysis that checks for their existence. In practice, trends in failure data often exist. Before the standard statistical methods can be applied, the nonexistence of statistical trends should be verified.

The realm of IID data is like the frictionless wheels, massless beams, and adiabatic heat transfer worlds of text book problems. All are ideal, but not realistic. The previous discussions point out the importance of checking for IID data properties, but did not describe how to perform such calculations. This section describes one such procedure. It is not unique or perfect, but it does address the key IID issues in a statistically meaningful way. It is useful for testing whether the data satisfy the property of being identically distributed. In practice, however, the procedure does much more. Trend analysis gives the user a visual display of the system failure history. The graphical, visual presentation of failure data assists in deciding if data sets do or do not exhibit trend behavior and follow a specific pattern not related to any specific type of trend but a pattern just the same.

A trend is defined by three properties:

1. Trend Existence Probability: the probability that a trend exists among data values over the given time interval.
2. Trend Type: Improvement or Deterioration.
3. Trend Strength: How fast reliability is changing.

The next sections describe some of the statistical information and tests that can be used to identify and measure the properties of trends. They contain both general information about the tests and specific details and formulas for performing the tests. Each reader should attack these sections at the level that will best serve his or her needs and background. They will supply an elementary understanding of the procedures and what is involved in using them, descriptions of when each should and should not be used, and information to help interpret their results. Use them as a basis for discussion, but do not expect to perform the tests based only on the summaries presented here.

Property #1: Trend Existence Probability

The trend procedure uses four separate test statistics to assist in the decision process regarding the existence or nonexistence of trends. They are Laplace [9-10], MIL-HDBK-189 [11], rank [12, 13], and linear regression [14]. The first two tests are robust, nonlinear methods designed for small data sets. They also use the valuable information of the time from the last failure to the end of the time period. They were designed with the added sensitivity that an improvement trend can be realized from *not* experiencing any failures for a sufficiently long period of time. The last two tests use only the actual data values to quantify the likelihood of trend existence. Let's see how these test statistics work.

The tests are discussed for the situation where a system is examined for a specified time. To the practitioner this may sound like a trivial statement. It is not. From a statistical perspective, this is a type of censorship. The test is influenced by an external constraint, a fixed period of time. The period of time is determined independent of the number of failures. In statistical jargon, this is called Type I censorship. The other type of censorship is when the test period is determined by the number of observed failures. This is called Type II censorship.

To present the test statistics, we need to describe the failure times in an analytical framework. During the time interval (0, T), suppose n failures have been observed. The successive time to failures are denoted by T_i, $i = 1$ to n, such that

$$0 < T_1 < T_2 < T_3 < \ldots < T_n < T$$

Laplace Test

The Laplace Test uses the failure data to compute a test statistic, U, that approximates a standard normal distribution.

$$U = \frac{\sum_{i=1}^{n} T_i - n\frac{T}{2}}{T\sqrt{\frac{n}{12}}}$$

(3-1)

where n = number of failures or events

   T_i = successive times to failure

   T = the total time period, $0 < T_1 < T_2 < T_3 < ... < T_n < T$

It has been shown that the Laplace Test result is adequate for n ≥ 4 at the 5% level of significance [15]. This last statement is very important. It says that the procedure can identify the existence or nonexistence of trends at the 95% level of significance with as few as four time-between-failure values or five actual failure dates. In practice, any trend identified with this small number of failures will probably be not be useful for accurate *predictions* of failures, but the knowledge of the existence or nonexistence of a trend is valuable information by itself.

The statistical procedure is demonstrated on the following two data sets that show 6 successive times-to-failure values. The analysis time interval was selected as one year (T = 365 days).

	Cumulative Time to Failure (days)					
System A:	36	57	73	102	105	205
System B:	160	260	263	292	308	329

Applying Equation 3-1 to System A, U = −2.00. Applying it to System B, U = + 2.00. From the normal distribution tables, these values indicate slightly over a 95% two-sided confidence level. It is not obvious from the above data sets that trends are indeed highly likely to exist. To show this behavior in a more intuitive way, we compute the time-between-failure values from the listed times to failure:

	Time Between Failures (days)				
System A:	21	16	29	3	100
System B:	100	3	29	16	21

Now System A can be seen to have an improvement trend behavior. The times between failures are, for the most part, getting larger. Also, there is a relatively large period of time, 165 days, from the last failure to the end of the time period. The interval can have an important influence on the likelihood for existence of a trend. An improvement trend can be obtained from a data set with many failures. As long as the times between successive failures are getting larger with each failure, an improvement trend will be indicated. However, an improvement can also be realized by having no failures for a relatively long period of time. The Laplace test checks data for this behavior.

MIL-HDBK-189

The MIL-HDBK-189 test statistic, V, is a variant of Laplace's test. It can be derived from the Laplace test using a slightly different set of assumptions.

$$V = 2 * \sum_{i=1}^{n} \ln \frac{T}{T_i} \tag{3-2}$$

where n = number of failures or events
T_i = successive time to failures or events
T = the total time period, $0 < T_1 < T_2 < T_3 < ... < T_n < T$

The test is applied much in the same manner as the Laplace test. The test statistic is computed from the successive times to failure values. The difference is that V is distributed as a chi-squared distribution (χ^2) with 2n degrees of freedom. Applying Equation 3-2 to the data sets for Systems A & B yields values for V of 17.8 and 4.0, respectively. Referring to the chi-squared distribution tables for 12 degrees of freedom, the test verifies the improvement and deterioration trends at approximately the same level of confidence as with the Laplace Test.

Both the Laplace Test and the MIL-HDBK-189 test are simple to apply and easy to calculate. In each case, one relatively straightforward formula is used to compute a test statistic. This value and, in the case of the MIL-HDBK-189 test, degrees of freedom are used to reference standard distribution tables and evaluate the probability that a trend exists. The failure data and the time from the last failure to the end of the time period are used explicitly in these calculations.

The next two tests use only the failure data and involve slightly more computation.

Rank Test

The Rank Test is derived from a procedure that makes no specific distribution assumptions regarding the mathematical form of the underlying processes. It is based solely on the rank order of the times between failure values. The procedure is demonstrated using the System A data set (Table 3-1. The chronological time between failure values are placed in column 1. The rank, which is the numerical number that corresponds to a particular value's size (smallest to largest) relative to the other values, is given in column 2. Column 3 contains each data value's score, $S_{r,m}$ as given by Equation 3-3. It is the expected value of the rth largest of m independent random variables following the exponential distribution with unit mean.

$$S_{1,n} = \frac{1}{n}$$

$$S_{2,n} = S_{1,n} - \frac{1}{n-1}$$

$$S_{r,n} = S_{r-1,n} + \frac{1}{n-r+1} \qquad (3\text{-}3)$$

$$S_{n,n} = S_{n-1,n} + 1$$

where n = total number of failures or events

r = an integer ranging from 1 to n

The last column shows the independent variable, z_i, as the usual linear orthogonal polynomials. The index, i, could be used, but this complicates the analysis. The linear orthogonal polynomials are constructed to have a zero mean. This simplifies the analysis. The polynomials can be easily constructed for any m. For example, if n = 6, the polynomials are −5, −3, −1, 1, 3, 5. For n = 7 the polynomials are −6, −4, −2, 0, 2, 4, 6.

The test statistic that is constructed from the data in Table 3-1 is asymptotically normally distributed with a mean equal to zero.

Table 3-1
Worksheet For Rank Test Calculation: System A

Time Between Failures	Rank	Score ($S_{r,n}$)	Independent Variable z_i
21	3	0.78	−4
16	2	0.45	−2
29	4	1.28	0
3	1	0.20	2
100	5	2.28	4

$$W = \sum_{i=1}^{n} (\text{ith score}) * z_i \qquad\qquad (3\text{-}4)$$

where n = total number of failures or events

z_i = a set orthogonal polynomials

Using Table 3-1 and Equation 3-4, $W = 5.5$. The distribution however, does not have unit variance and, therefore, cannot be directly used to access the standard normal distribution tables. To correct this, W must be divided by the distribution standard deviation that is given by

$$s = \sqrt{K_{2,n} * \sum_{i=1}^{n} z_i^2}$$

where $K_{2,n} = 1 - \dfrac{1}{n-1} * \left(\dfrac{1}{n} + \dfrac{1}{n-1} + K + \dfrac{1}{2} \right)$.

The standard normally distributed test statistic is:

$$u = \frac{W}{s} \qquad\qquad (3\text{-}5)$$

For this example, $K_{2,5} = 0.68$ and the sum of squares, computed from data in Table 3-1, is 40. Multiplied together and applying the indicated square root yields the standard deviation of 5.21. The standardized normal statistic is computed using Equation 3-5 as $u = 1.06$. Comparing this result with the Laplace test statistic ($u = 2.00$) shows that an improvement trend is observed at a lower level of confidence. From the standard normal distribution tables, $u = 1.06$ indicates a confidence level of approximately 85%, compared to over 95% computed from the Laplace test.

Linear Regression

The Linear Regression test involves fitting the linear regression to the failure data. The form of the equation is

$$y(i) = m * i + b \qquad\qquad (3\text{-}6)$$

where $y(i)$ = the time between failures predicted from the regression equation

m = the slope of the straight line

$m > 0$ suggests an improvement trend

$m < 0$ suggests a deterioration trend

$m = 0$ suggests no trend

i = the failure number ($i = 1,2,3,..., n$)

b = value where the line intersects the time axis

As can be seen, trend information comes from the slope, m. The test develops the probability that the failure data slope has the sign of the calculated mean value. In other words, it shows the probability that an improvement or deterioration trend exists. The slope, m, and intercept, b, are calculated from the failure data by the following equations:

$$m = \frac{\sum_{i=1}^{n} i * y_i - n * \bar{i} * \bar{y}}{\sum_{i=1}^{n} i^2 - n * \bar{i}^2}, \, b = \bar{y} - m * \bar{i} \tag{3-7}$$

where n = total number of failures

y_i = the time from the beginning of the recording period to the ith failure

i = the failure number ($i = 1,2,3,..., n$)

$\bar{y}$ = arithmetic mean of the failure times

$\bar{i}$ = mean failure number, $\dfrac{1}{n} \sum_{i=1}^{n} i$

Confidence limits span an interval that has the slope value, computed by Equation 3-7 as its center. Given any failure data set, the range of this interval depends upon the desired significance level. The larger the significance, the larger the range. If m_0 denotes the actual, unknown slope, then the uncertainty range defined by m, the confidence level, $f(\alpha)$, and the data dependent functions $S_{y,x}$. S_x is given by

$$m - f(\alpha) \frac{S_{y,x}}{S_x} < m_0 < m + f(\alpha) \frac{S_{y,x}}{S_x} \tag{3-8}$$

where $f(\alpha) = \dfrac{t_\alpha}{\sqrt{n-2}}$

t_α = the value from the t distribution corresponding to a prescribed significance level, a, and n − 2 degrees of freedom

$$S_{y,x} = \sqrt{\sum_{i=1}^{n}(y_i - y(i))^2}, \qquad S_x = \sqrt{\sum_{i=1}^{n}(i - \bar{i})^2}$$

y_i = the actual failure data values

$y(i)$ = the values predicted from the regression equation computed from Equation 3-6

These formulas show how to compute the linear regression parameters and confidence interval given a desired significance level. The procedure to compute the probability that an improvement or deterioration exists uses Equation 3-8 as a starting point.

The data values determine all of the parameters except the confidence level, $f(\alpha)$. This value is determined from the desired significance level of the analyst. For trend existence, however, we are more interested in the sign of the slope than its actual value. A positive slope indicates an improvement trend, a negative value indicates a deterioration trend, and a zero value signifies neither improvement nor deterioration. Thus, we are interested here in the value of $f(\alpha)$, and in particular, the value of t that causes the range to change sign.

This is best described with an example. Suppose that for a given problem where n = 11, Equation 3-8 is computed by

$$2 - \frac{4t_\alpha}{3} < m < 2 + \frac{4t_\alpha}{3} \tag{3-9}$$

Because the calculated slope value, 2, is positive, the value of t_α that will extend the range to a negative number must be determined. Referring to t-distribution tables found in statistics texts, with this value at the n-2 or 9 degrees of freedom identifies the level of confidence or the probability that the real slope value is positive. This value also indicates the probability that an improvement trend exists. To perform this calculation, the left side of Equation 3-9 is set equal to zero and solved for t_α. The value is 1.5. Now, referring to the t-distribution tables with 9 degrees of freedom, the indicated level of confidence is approximately 85%. From the linear regression test, there is a 0.85 probability that the failure data indicates an improvement trend. If the computed slope is negative, indicating a deterioration trend, then the right side limit is used as shown above to determine the probability that the slope value is indeed negative.

The linear regression test is now applied to system A of the same example that we used for the other tests. The calculation's intermediate and final results are as follows:

Linear Regression Trend Test

$b = -9.7$ Linear regression equation: $y(i) = 14.5*i - 9.7$
$m = 14.5$ Note the positive slope implies an improvement trend.
$S_{y,x} = 61$ Degrees of freedom: $n - 2 = 3$
$S_x = 3.2$

Equation 3-9 can be written as

$$14.5 - \frac{61t_\alpha}{3.2\sqrt{3}} < m < 14.5 + \frac{61t_\alpha}{3.2\sqrt{3}}$$

Setting the left side limit equal to zero and the resulting equation is solved for t_a.

$$14.5 - \frac{61t_\alpha}{3.2\sqrt{3}} = 0$$

This process yields $t_\alpha = 1.32$. Referring to the t distribution tables for 3 degrees of freedom shows the probability that the slope is positive is roughly 0.75. This indicates that actual improvement is about 75% likely.

Statistical Test Limitations

Each of the analysis techniques previously described is derived under specific assumptions and has accuracy limitations in its application. The Laplace is not formally derived under any specific failure model, but it is a good test if the rate of occurrence of failure (better known as the failure rate), $\rho(t)$ is known to follow the function given in Equation 3-10 [15].

$$\rho(t) = e^{a + bt}, \quad -\infty < a, b < \infty \tag{3-10}$$

The MIL-HDBK-189 tests assumes that the underlying actions governing failure occurrences together form a Poisson process with a rate of occurrence of failure that varies in time by the following related form [16].

$$\rho(t) = \lambda\beta t^{\beta-1}, 0 < \lambda, \beta < \infty \tag{3-11}$$

In both of these equations the constants are determined from formulas involving the failure data. Equations for these parameters can be found in the referenced literature. Equations 3-10 and 3-11 present some of the underlying assumptions involved with using the Laplace and MIL-HDBK-189 test statistics.

Equations 3-10 and 3-11 describe what is called in the reliability literature as Non-homogeneous Poisson Processes, or NHPP for short. The usual Poisson distribution is a probability distribution model with only one parameter. In other words, the mathematical distribution function is completely defined by one parameter. Processes that are modeled by this version are called Homogeneous Poisson Processes or HPP. The NHPP is derived from the HPP by replacing its usually time independent parameter with a time dependent function such as Equation 3-10 or Equation 3-11.

Equation 3-11 is a fairly famous form that has received a lot of attention. Processes that are modeled with this failure occurrence pattern are called "weibull processes." If β is determined to be greater than 1, the time between failures is getting smaller, indicating that performance is deteriorating. If β is less than 1, the time between failures is getting larger, indicating improvement of system performance. No trend, or a flat trend is given by $\beta = 1$. In this case, the failure occurrence rate is constant. Equation 3-11 will be used in the next section to develop an estimate for future failure time predictions.

The name is somewhat confusing because situations that are described by the weibull process form are different than situations modeled with the weibull probability distribution. Although the two theoretical approaches share some common ground, *they are very different methods*. There is a general confusion between these two concepts—be careful. The weibull probability distribution is a stationary distribution that requires IID data. A weibull process does not produce IID data by the fact that the data are not identically distributed.

How general is the Poisson form? The good news here is that the Poisson distribution is a very general probability model that can be accurately applied to an extremely wide range of situations. If there was one model that could be chosen for a general tool over all others, it would be the Poisson. We are fortunate that nature allows us this mathematical and practical convenience. Remember, however, that the Poisson version does not describe all possible situations in the real world.

The Laplace and MIL-HDBK-189 tests account for the possibility that the equipment is changing in time and the stochastic processes governing failure occurrence are time dependent. The MIL-HDBK-189 test assumes that the failure occurrence rate is given specifically by Equation 3-11. The Laplace test method does not pre-suppose the existence of any specific formula. It only accounts for the possibility that the rate of failure occurrence is changing in time. The tests should be used with care because their validity or accuracy are dependent on the rate of failure occurrence mathematical form. Use them with all of the other information you can compile to make conclusions regarding changes in reliability growth.

The Rank test also does not assume a specific type of process is involved in determining the occurrence of failure events. It assumes the existence of some undetermined trend-free form. The only real potentially restrictive assumption is that the failure events are independent. As discussed earlier, this is not always possible due to insufficient data. In any case, because of the possibility of synergistic influences between data values, the best technique for judging failure event independence is a comprehensive review of the data by people familiar with the systems.

The Linear Regression test involves a purely mathematical fit of a straight line to the failure data. The confidence interval expression for the slope is used to compute the maximum probability that the sign of the slope does not change. This procedure is the least appealing from an intuitive point of view. Straight line, least-square-curve fitting would seem to be an inaccurate method of modeling dynamic changes in processes. When there are sufficient data, this method works pretty well. An explanation for its success is that we are not asking the procedure for a great deal of information. The linear regression formalism is used only to produce the probability that the calculated slope value is of the correct sign. From a mathematical point of view, this is very non-specific information. In general, accuracy of Linear Regression testing grows as the amount of data increases.

Trend Existence Results

The results of these four tests are displayed as a 4-tuple, representing the percent probability that each test identifies a trend among the data values. The four tests are like a set of experts, each with a particular talent to identify specific types of trend behavior. They all examine the data and then separately give the user their opinion on whether a trend exists or not. In essence, each of the tests—Laplace, MIL-HDBK-189, Rank,

Linear Regression—are designed to recognize certain types of patterns in failure data.

The standard statistical inference procedure could be applied here. A level of significance could be set and trend existence could be tested at this level. From a practical perspective, this type of analysis is very inefficient. Here's an example that shows why. Suppose the level of significance set at 0.05. A test was performed and the test statistic produced a probability of existence of 90%. Standard statistical inference procedures say that no trend exists at the 95% level of confidence. However, to practitioners, the 90% level may be fine! Thus, the percent probability that each test identifies a trend among the data values is explicitly given. This information empowers the practitioners to decide for themselves whether or not a trend exists. Remember, statistics can not be wrong. There are always significance levels and assumptions. It makes sense to give decision makers the maximum amount of information.

Property #2: Trend Type

Understanding the type of trend is a simple but very important aspect of the trend analysis. This property describes whether the system is improving or deteriorating over the prescribed time interval. Because I included these trend types in the discussions of trend existence, I won't go into them any further here.

Oscillations in failure behavior are not explicitly analyzed. Techniques designed to recognize oscillating relationships require more data than usually available in failure databases.

Property #3: Trend Strength

Property #1 gave the probability of the existence of a trend. Property #2 indicated what type of trend may exist. This information is necessary, but it is incomplete. It is not enough information to adequately classify improvement and deterioration trends. Suppose a system is analyzed separately for two separate years. Properties #1 & #2 indicate that improvement trends have a high probability of existence in both years. How can the results be compared? Without some measure of how rapidly (or slowly) reliability is changing, there is no way to distinguish between them. For example, an almost flat improvement trend could exist in the first year while the second year data could result in a steep, much greater, improvement trend.

The strength of a trend indicates how fast or slow reliability is changing. It is computed by comparing the mean time between failures (MTBF) and the predicted time to next failure (PTNF) calculated from the observed trend analysis. If these two numbers are much different, then the observed trend is steep. A shallow trend is suggested when the two numbers are about the same. A completely flat trend occurs if the MTBF and the PTNF are exactly equal.

Predicted Time to Next Failure (PTNF)

As just discussed, the MIL-HDBK-189 test for trends assumes that the rate of occurrence of failures has the form given by Equation 3-11. The constants, λ and β, can be determined by the maximum likelihood method in terms of the data values. The time to successive failures is indicated by $\{T_i\}$. The details of this very powerful statistical technique are not appropriate for this book. It is sufficient to say that for large data sets, the maximum likelihood method provides an accurate way to determine parameters for probability-related quantities. For small data sets where accuracy is really not possible, it is usually the best way to calculate probability-related parameters when decisions need to be made.

Equation 3-11 is rewritten here for easy reference. Its functional form indicates the rate of occurrence of failures, $\rho(t)$, and is described by a simple model that varies as time is raised to some power:

$$\rho(t) = \lambda\beta t^{\beta-1} \tag{3-11}$$

The maximum likelihood fit to the data yields the following formulas for λ and β:

$$\beta = \frac{n}{n\ln(T) - \sum_{i=1}^{n}\ln(T_i)}, \qquad \lambda = \frac{n}{T^{\beta}}$$

where n = number of failures or events

T_i = successive time to failures or events

T = the total time period, $0 < T_1 < T_2 < T_3 < ... < T_n < T$

In terms of the MIL-HDBK-189 test statistic, and the total number of failures, n,

$$\beta = \frac{2n}{V}$$

For small sample sizes, it is advisable to use an unbiased estimate given by

$$\beta_{ub} = \frac{n-1}{n}\beta$$

The correction factor of $(n-1)/n$ ensures statistically that the sample mean of the estimate used for β agrees with the mean of the ideal, population value in the limit as $n \to \infty$.

The question is, which version of β should you use in actual computations? In general, the unbiased version, β_{ub} seems to be more accurate. As the size of the failure data set grows, the difference between the two forms becomes less important. β_{ub}, is used exclusively for all calculations in the text examples.

The assumption that Equation 3-11 does indeed adequately fit the failure data can also be verified at a given significance level. In statistical terms, this corresponds to testing the null hypothesis that a non-homogeneous Poisson process with the failure occurrence function given by Equation 3-11 describes the reliability growth of the particular system under study. This is accomplished by using the Cramer-von Mises statistic, C^2_n. For the case discussed in this text, where n failures were observed in the time interval $(0,T)$,

$$C_n^2 = \frac{1}{12n} + \sum_{i=1}^{n}\left(\left(\frac{T_i}{T}\right)^{\beta_{ub}} - \frac{2i-1}{2n}\right)^2 \tag{3-12}$$

where   n = number of failures or events
        i = an integer counter going from 1 to n
        T_i = successive time to failures or events
        T = the total time period: $0 < T_1 < T_2 < T_3 < ... < T_n < T$

The null hypothesis is rejected if the statistic, C^2_n, exceeds the critical value for the user selected level of significance. Critical values of C^2_n are given in Table 3-2 for the significance levels: 0.20, 0.15, 0.10, 0.05, and 0.01. This model of reliability growth is very flexible and can be applied to a broad range of problems. If a particular data set does not pass this test

Table 3-2
Critical Values for the Cramer-von Mises Goodness-of-Fit Test

	Level of Significance: α (Degree in Confidence)				
n	0.20 (80%)	0.15 (85%)	0.10 (90%)	0.05 (95%)	0.01 99%)
2	0.138	0.149	0.162	0.175	0.186
3	0.121	0.135	0.154	0.184	0.230
4	0.121	0.134	0.155	0.191	0.280
5	0.121	0.137	0.160	0.199	0.300
6	0.123	0.139	0.162	0.204	0.310
7	0.124	0.140	0.165	0.208	0.320
8	0.124	0.141	0.167	0.210	0.320
9	0.125	0.142	0.167	0.212	0.320
10	0.125	0.142	0.169	0.212	0.320
11	0.126	0.143	0.169	0.214	0.320
12	0.126	0.144	0.169	0.214	0.320
13	0.126	0.144	0.169	0.214	0.330
14	0.126	0.144	0.169	0.214	0.330
15	0.126	0.144	0.169	0.215	0.330
16	0.127	0.145	0.171	0.216	0.330
17	0.127	0.145	0.171	0.217	0.330
18	0.127	0.146	0.171	0.217	0.330
19	0.127	0.146	0.171	0.217	0.330
20	0.128	0.146	0.172	0.217	0.330
30	0.128	0.146	0.172	0.218	0.330
60	0.128	0.147	0.173	0.220	0.330
100	0.129	0.147	0.173	0.220	0.340

For n > 100, use n = 100 values.

at the desired level of significance, there are two possibilities you can explore to correct this. First, examine the data to see if more than one failure occurs within the same time frame. If so, there is another procedure to use that we won't delve into at this point [16]. Second, check the data values for any obvious discontinuities or abrupt, sustained changes. If any breaks (discontinuities) are found, then the data set should be divided into separate groups and analyzed individually [17].

Systems A and B data sets are used to demonstrate the Cramer-von Mises test and to show the numerical behavior of λ and β for improvement and deterioration trends. The successive times to failure for both systems are presented again below for convenience:

System	Cumulative Time to Failure (days)					
System A:	36	57	73	102	105	205
System B:	160	260	263	292	308	329

The total time period is one year, $T = 365$. The results are given in Table 3-3.

Table 3-3
Reliability Growth Model Parameters & Systems A & B
Cramer-von Mises Test Statistics

	System A	System B
β	0.676	3.017
β_{ub}	0.563	2.51
λ	0.217	2.2×10^{-6}
C^2_6	0.171	0.078

System A shows an improvement trend, with $\beta < 1$. System B shows a deterioration trend, with $\beta > 1$. The Cramer-von Mises test verifies the applicability of the reliability growth model, Equation 3-12 in both cases. In Table 3-2, look at the row for $n = 6$. System A's test value of 0.171 indicates the level of significance is approximately 0.075 or the degree of confidence of 93%. For System B, the deterioration trend data fit to the model parameters yields $C^2_6 = 0.078$. This value is less than all critical values, which suggests that the model is acceptable.

The Predicted Time To Next Failure

Equation 3-11 is used in the following way to predict the time to next failure at any time, t. It can be integrated over time to calculate the expected number of failures between t and t + Δt.

$$E[N(t + \Delta t) - N(t)] = \int_t^{t+\Delta t} \lambda \beta \tau^{(\beta - 1)} d\tau = \lambda t^{\beta} \Big|_t^{t+\Delta t}$$

Because we know the data set has n failures, the expected number of failures at time t_n is equal to n. In mathematical terms:

$$E[N(t_n) - N(0)] = \int_0^{t_n} \lambda\beta\tau^{(\beta-1)}d\tau = \lambda t^\beta \Big|_0^{t_n} = n$$

To predict the cumulative time to the next failure, t_{n+1}, we set the expected number of failures to occur during the interval of t_n to t_{n+1} as 1.

$$E[N(t_{n+1}) - N(t_n)] = \int_{t_n}^{t_{n+1}} \lambda\beta\tau^{(\beta-1)}d\tau = \lambda t^\beta \Big|_{t_n}^{t_{n+1}} = 1$$

Solving for t_{n+1} gives the predicted time to the next failure.

$$t_{n+1} = \left(t_n^\beta + \frac{1}{\lambda} \right)^{\frac{1}{\beta}} \tag{3-13}$$

This time is the next successive, cumulative time of the next failure. What is usually desired, however, is the time from the last failure to the next (future) event. This is given by subtracting the time of the last failure, t_n, from the result given by Equation 3-13.

This prediction of the future uses all of the information contained within the data set. The prediction is useful in obtaining an indication of what the trend suggests as the next failure time. Another practical use of the prediction is to compare the predicted value with the MTBF to obtain an idea of the potential steepness or strength of the trend.

The Mean Time Between Failure Connection

The parameter, λ, is related to a version of the mean time between failures (MTBF). The difference now is that this version of the MTBF is a function of time. The instantaneous mean time between failures, IMTBF, is given by

$$IMTBF(t) = \frac{1}{\rho(t)} = \frac{1}{\lambda\beta t^{\beta-1}} \tag{3-14}$$

In practice, this provides an easily calculated estimate for the predicted time to next failure from the end of the time period. The PTNF computed in the last section and the IMTBF usually give about the same results. The advantage of using the IMTBF is that approximate confidence intervals can be easily computed.

Depending on the type of trend specified by the value of β, the instantaneous MTBF becomes smaller or larger as time progresses. Also notice that for a flat or no trend case where $\beta = 1$, the IMTBF loses its explicit time dependence. For improvement trends, $\beta < 1$. For flat or no trend, $\beta = 1$. For deterioration trends, $\beta > 1$.

Confidence interval estimation has been determined for our situation where we terminate the test period at some prescribed time, rather than by some predetermined number of failures. The probability distribution of the IMTBF point estimate computed for $t = T$, the end of the time period, is the basis for interval determination. Table 3-4 provides two-sided interval estimates for the ratio of the true IMTBF to the estimated IMTBF for several values of sample size. This ratio can be manipulated to yield approximate confidence intervals.

Table 3-4
Confidence Parameters for IMTBF Interval
Confidence Coefficient γ

N	$\gamma = 0.80$		$\gamma = 0.90$		$\gamma = 0.95$		$\gamma = 0.98$	
	$L_{N..80}$	$U_{N..80}$	$L_{N..90}$	$U_{N..90}$	$L_{N..95}$	$U_{N..95}$	$L_{N..98}$	$N..98$
2	0.261	18.66	0.200	38.66	0.159	78.66	0.124	198.7
5	0.426	3.386	0.352	4.517	0.300	5.862	0.250	8.043
10	0.549	2.136	0.476	2.575	0.421	3.042	0.367	3.712
15	0.614	1.800	0.545	2.087	0.492	2.379	0.438	2.781
20	0.657	1.638	0.591	1.858	0.540	2.076	0.488	2.369
25	0.687	1.540	0.625	1.722	0.576	1.900	0.525	2.134
30	0.711	1.475	0.651	1.631	0.604	1.783	0.554	1.980
35	0.729	1.427	0.672	1.565	0.627	1.699	0.579	1.870
40	0.745	1.390	0.690	1.515	0.646	1.635	0.599	1.788
45	0.758	1.361	0.705	1.476	0.662	1.585	0.617	1.723
50	0.769	1.337	0.718	1.443	0.676	1.544	0.632	1.671
60	0.787	1.300	0.739	1.393	0.700	1.481	0.657	1.591
70	0.801	1.272	0.756	1.356	0.718	1.435	0.678	1.533
80	0.813	1.251	0.769	1.328	0.734	1.399	0.695	1.488
100	0.831	1.219	0.791	1.286	0.758	1.347	0.722	1.423

$$L_{n,\gamma} = \left(1 + z_{0.5+\gamma/2} / \sqrt{2n}\right)^{-2} \qquad U_{n,\gamma} = \left(1 - z_{0.5+\gamma/2} / \sqrt{2n}\right)^{-2}$$

Given a sample size, n, and the selected confidence coefficient, the table values L_n,γ and U_n,γ can be selected from Table 3-4 or computed from the formulas given for n > 100. The interval for the IMTBF is given by

$$L_n,\gamma * IMTBF(t) \leq IMTBF \leq U_n,\gamma * IMTBF(t) \tag{3-15}$$

Additional table values can be found in Reference 18. This is not the last word on the subject of confidence intervals for reliability growth results. The knowledge base is still growing [19].

To illustrate how the observed statistical trend relationship can affect the MTBF value, the instantaneous IMTBF, computed for t = 365, the end of the time period, and the standard MTBF are given below for Systems A & B. A 90% confidence interval is computed for the MTBF using standard descriptive statistics methods, and a 90% confidence interval for the IMTBF is computed using Equation 3-15 and the values given in Table 3-5.

Table 3-5
Comparison of IMTBF & MTBF 90% Confidence Intervals for Improving and Deteriorating Systems

System A
IMTBF (365) point estimate: 108 days 90% Confidence Interval: [38,488] MTBF point estimate: 34 days 90% Confidence Interval: [0, 72]
System B
IMTBF(365) point estimate: 24 days 90% Confidence Interval: [9,109] MTBF point estimate: 34 days 90% Confidence Interval: [0, 72]

There is an important fact illustrated by the previous example. The standard MTBF equation *cannot identify the presence of time dependent behavior* in data. To compute the standard MTBF data values, the times between failures are summed and divided by the number of points. The chronological order of the data values is not considered in the computation. Thus, as previously seen, regardless of whether an improvement or deterioration trend is present, the MTBF value is the same. *The MTBF only becomes applicable when no trend is verified to exist in the failure data.*

Also note that the confidence intervals for the IMTBF are not symmetric around the IMTBF point estimates. This is caused by the inherent time dependence of the failure rate form given in Equation 3-14. These confidence intervals are larger than normally seen in standard statistics. The factors, L_n,γ and U_n,γ are relatively large. Thus, in practice, they may or may not be useful. The point is that the IMTBF can be used more as a rough indicator of failure time than to predict the future. Confidence intervals can give the illusion of genuine confidence in the results. In analyzing failure data, the IMTBF is just one more piece of information to assist in the decision process. Because the confidence intervals for the IMTBF are usually large, should they be computed along with the point estimate? Yes, the additional calculation effort is small, and statistically it's the right thing to do.

Trend Analysis Examples

Four examples, showing an Improvement Trend, a Deterioration Trend, No Trend, and a Data Pattern are now presented to illustrate the principles of statistical trend identification and characterization. The results of the four statistical trend tests are given as a 4-tuple in the order, Laplace, MIL-HDBK-189, Rank, and Linear Regression. The trend type (improvement or deterioration) and the trend strength are given above each plot to show the results of the statistical analysis of trend identification and characterization.

The graphs show the time between failure data points and several statistical values designed to help the user decide whether or not a trend exists. These plots are extremely valuable in the trend analysis procedure. The failure data plots showing how the time between failures changes with failure number provides a historical view of the performance of the component or system. In many cases, people familiar with the processes under study will look at these plots and be able to describe the circumstances involved with the various failure events. The visualization of the failure data in this form is the first step in analyzing why a trend appears to be occurring. The analytical statistical tools offer additional evidence to assist in the trend decision process, but they are not designed or intended to *be* the decision process. This is why the plots are so important. They give the user a visual tool that can be combined with plant experience to decide if a trend really exists. The trend statistics by themselves do not consider the normal synergy involved with plant operations and maintenance.

All of the graphs have the same format. The dotted line shows the time behavior of the MTBF over the failure history. It is visual evidence to use in deciding if the MTBF is a representative measure of future reliability.

The straight line is a linear regression fit to the data points that are denoted by the connected squares.

The regression curve is another visual aid to show the overall behavior of the time between failures. The arrow highlights the prediction estimates of the time to the next failure. The cross is the predicted failure time computed using Equation 3-14. The linear regression prediction is obtained by extrapolating to the n+1th failure.

Example #1: An Improvement Trend

The first example illustrates failure data compiled over a time period for a system of pumps. Failures from all effects, e.g., seals, bearings, etc., were combined into one data set. The analysis is given in Figure 3-2.

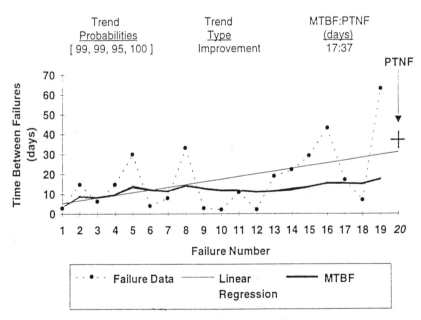

Figure 3-2. Improvement trend example. (Reprinted with permission from *CheckMaint!*™)

Figure 3-2 shows that statistical improvement does not necessarily mean simple, linearly increasing times between failures. By looking at Figure 3-2 you may or may not agree that the failure times are, in general, getting larger. Remember, the improvement conclusion uses the data you can see as well as one fact that is not apparent from the data, the time from the last failure to the end of the time period. All of this information says an improvement trend does exist. In fact, the strength of the specific trend is fairly steep, 17:37. The trend indicates that the time from the last failure to the next future event is more than twice the standard MTBF. Linear regression, which uses only the failure data values in computing its trend results, also indicates the next failure is significantly larger than the MTBF. Notice how slowly the MTBF changes in time. This behavior is characteristic of the mean. As the number of data values increases, its sensitivity to change from new, different values decreases rapidly. Usually within three to five data values the MTBF is generally determined and from this point forward changes very slowly. All of the examples here illustrate this point. The MTBF by design and in practice is a time-independent statistic.

Example #2: A Deterioration Trend

Failure data was compiled for a compressor system, for a 4-year 10-month time period. The analysis results are given in Figure 3-3.

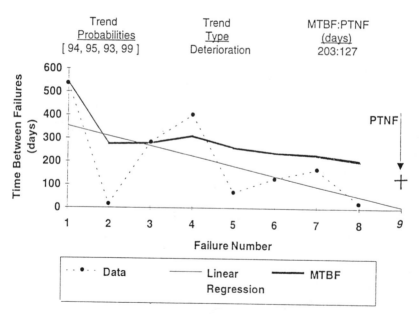

Figure 3-3. Deterioration trend example. (Reprinted with permission from *CheckMaint!*™)

The deterioration behavior of the data is apparent from the plot of Figure 3-3. However, it is conceivable that the deterioration trend would not be noticed in practice. With only eight failures in almost five years, system failures have such a low frequency that the time relationship could go unnoticed compared to the always present higher priority issues. In this case, the MTBF is considerably larger than all of the PTNFs. This is not surprising because the MTBF was initially computed as a relatively large value and its subsequent change from the addition of one more data point is small. The deterioration trend strength (207:127) suggests that the next failure will occur at a time less than the MTBF.

Example #3: No Trend

Failure data was compiled for a three-cell cooling tower over a nine-year time period. The tower consists of three fixed-bladed propellers, three gear boxes, three fan motors, three water pumps, two pump motors, and one steam turbine. A failure was recorded for this system when any one or more of these components failed. The failure history and trend results are shown in Figure 3-4.

The statistical trend results were mixed. The Laplace and MIL-HDBK-189 tests indicated relatively low probability improvement trends as indi-

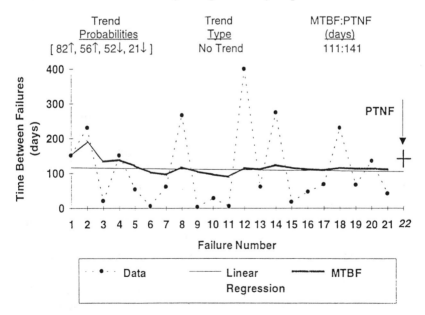

Figure 3-4. No trend example. (Reprinted with permission from *CheckMaint!*™)

cated by the arrow ($\uparrow$). The other two tests gave low probabilities of deterioration trends, indicated by the downward arrow ($\downarrow$). The practical conclusion drawn from these mixed results is there is no trend in the data. The reliability of the cooling tower is not getting any worse or better. In this case, the MTBF is a good candidate for the representative time between failures. In principle, the data values should be tested for independence. As mentioned earlier in the chapter, statistics may not be the best way of doing this. People familiar with the system over this time period are probably better judges of component failure dependencies than mathematical procedures.

Example #4: A Data Pattern

This example was fabricated to show a particular danger of depending too much on just the trend statistics. When performing many analyses, it is easy to allow the statistics to decide for you instead of looking at each time between failure plot individually to verify what the numbers are saying. This example shows a very clear pattern in how often the system fails. It is not as uncommon as you might think to see deterministic failure patterns in failure data. This is true especially for analysis of large complex systems. In this example, the pattern is simple in construction. The plot and trend results are given in Figure 3-5.

The trend results suggest an improvement trend exists at very low probability levels. Without looking at the time between failures plot, the most logical conclusion is that no trend exists. This is true, there is no improvement or deterioration in the standard way, but the pattern is not recognized by the statistics. This should not be surprising. The statistics were not developed to recognize these patterns. Pattern recognition in industrial failure data often becomes more an art than a science. In practice, many failure data patterns are hard to see, and it is sometimes unclear whether or not the particular pattern is by chance or because of some underlying factors. In any case, the trend statistics do not tell the whole story; you must look at the failure data plots.

Patterns in failure data may be more complex than repeating shapes in failure graphs. For example, maintenance induced failures are seen by a repetition of a stochastic failure followed shortly by another failure. This pattern can be recognized if every other time between failure data value is relatively small. For example, suppose a motor-operated valve fails due to excessive wear. We can consider this a stochastic event. The maintenance technician fixes the problem, and two days later it fails again due to lack of adjustment. The second failure is maintenance-induced and would be

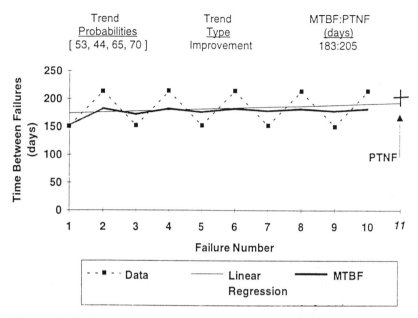

Figure 3-5. Data pattern example. (Reprinted with permission from *CheckMaint!*™)

indicated on the time between failure plot by a small value after a large value. You cannot expect to identify all maintenance failures this way, but if the pattern is suggested by the plot, further investigation is justified. Often the times between failure plots raise more questions than they answer. They provide insights into actual system performance that really cannot be obtained in any other way.

Summary

Statistics are powerful, effective tools to assist the management decision process. Their responsible application is more an art than a science. That is why it is not uncommon to spend more time in the design of statistical experiments than in the actual computations. Understanding their strengths and weakness in a particular application is essential if the study results are to be meaningful. Statistics are not "numbers looking for an argument." They're only looking for someone who can interpret their meaning correctly.

References

1. Smith, J. M., *Mathematical Modeling & Digital Simulation for Engineers & Scientists,* John Wiley & Sons, New York (1977).
2. Harmon, M. D., "When Figures Lie," Portland Press Herald, September 6, 1993, p 7A.
3. *Mademoiselle,* October 1993, p. 48.
4. Gasink, W. A., *Avoiding Statspeak,* JAWCO, East Stroudsburg, PA, 1991.
5. Gasink, op. cit. pp. 10–12.
6. Lutz, W., *Doublespeak,* Harper & Row, New York, NY, 1989, p. 43.
7. Vogas, J. L., "Sneak Analysis of Process Control Systems," First International Conference on Improving Reliability in Petroleum Refineries and Chemical and Natural Gas Plants, Houston, TX, Gulf Publishing Company, November, 1992.
8. Savakoor, D. S., Bowles, J. B., and Bonnell, R. D., "Combining Sneak Circuit Analysis and Failure Modes and Effects Analysis," ARMS (Annual Reliability and Maintainability Symposium), 1993, pp. 199–205.
9. Cox, D. R. and Lewis, P. A., *The Statistical Analysis of Series of Events,* Methuen, London, pp. 45–51, 1966.
10. Ascher, H. and Feingold, H., *Repairable Systems Reliability,* Marcel Dekker, Inc. New York, 1984, pp. 78–79.
11. MIL-HDBK-189, "Reliability Growth Management," Headquarters, U.S. Army Communications Research and Development Command, DRDCO-PT, Fort Monmouth, NJ 07702, 1981, p. 68.
12. Ascher and Feingold, op. cit. pp. 80–82,
13. Cox and Lewis, op cit. pp. 54–58.
14. Jones, R. B., *CheckMaint!* Manual, unpublished, 1990.
15. Bates, G. E., "Joint Distributions of Time Intervals for the Occurrence of Successive Accidents in a Generalized Polya Scheme," *Ann. Math. Stat.,* 26, (1955), pp. 705–720.
16. Goel, A. L. and Okumoto, K. M., "Time Dependent Error Detection Rate Model for Software Reliability and Other Performance Measures", *IEEE Transactions on Reliability,* R 28, no. 3, 1979, pp. 206–211.
17. Crow, L. H., "Reliability Analysis for Complex Repairable Systems," Reliability and Biometry-SIAM Philadelphia, 1974, pp. 379–410.
18. MIL-HDBK-189, "Reliability Growth Management," Headquarters, U.S. Army Communications Research and Development Command, DRDCO-PT, Fort Monmouth, NJ 07702, 1981, p. 143.
19. Crow, L. H., "Confidence Intervals on the Reliability of Repairable Systems," ARMS, 1993, pp. 126–133.

CHAPTER 4

Common Measurement Parameters and Applications

Man is the measure of all things.

Protagoras, (c.485–c. 410 B.C.)

Let's begin our discussion of measurement parameters from a philosophical point of view. Measures by themselves have no meaning. They are hollow definitions yielding no information until they are applied to realistic situations and interpreted properly. The quality and value of any measurement are directly related to the ingenuity applied to selecting the data boundaries or, in other words, how the "system" being measured is defined. Just as important are the ways the measurements, once defined and collected, are used.

Because the definition of the data boundaries is the first crucial step in successful measurement, we'll take some time now to discuss the meaning of systems and the hierarchy that is useful in describing combinations of them.

Systems and Super-Systems

Let's start with something that is commonplace, that is the notion of a system. This term is a familiar part of our language—whether it's used in electric shaver commercials, diet advertisements, or the Sunday comics. We see the word "system" everywhere. It would be an interesting study to plot the number of times the word was used in a major newspaper as a function of time. My belief is that not much change would be seen until about 1970, when there would a rapid increase in use of the word. This is probably a result of the increase in number of computers and other technological advances in our society.

The concept of a system is not firmly delineated by the laws of nature. It is a more abstract quantity, applied by us to assist in the description, identification, and application of some entity. Its utility is in the eye of the viewer. Some systems are suggested to us by their natural configuration, but still it is up to us to explicitly define them and their boundaries. In the broadest terms, *a system is a collection of things that together performs one or more functions.* Notice the two key words in the previous sentence: *collection* and *function.* Regardless of what kind of system is under discussion, it will always be a collection of components that perform a function or functions. The function (or functions) performed by the system cannot be performed by the components themselves. To apply a well-known phrase, the whole (collection) is greater than the sum of the parts (components).

The human body contains many systems. Each can be viewed as a collection of things that together perform many functions. The functions could not be performed by the parts by themselves. The combination of system components create and support the functions of the whole.

Another example of a system can be developed by looking at an automobile from a functional perspective. To most easily identify its systems you should first think about the functions that are performed to make the automobile work. Generally, these functions will define the systems. For example, consider the following short list of functions performed within an automobile: heating, cooling, transmission of energy, power, and braking. These functions are actually performed by various systems in the vehicle. It makes sense to identify systems by their primary functions because these are what make them unique. Each system also is an appropriate way of referring to a specific collection of automobile parts.

Another aspect of systems is the characteristic of a system boundary. This will be discussed at length in Chapter 5, but here we will mention it briefly. Every system has a unique, identifiable boundary that denotes its limits. System boundaries are often physical in nature and therefore can be defined very clearly. With the concept of system limits identified, we can now go on and give special meaning to what crosses or passes through these hypothetical walls. We call whatever passes across system boundaries its inputs and outputs. In physical systems, this structure of system boundaries can be used to model the specific functions and keep track of the details of quantities passing in and out of the systems. These quantities define the interfaces between systems.

In current applications of the term system, there is a secondary, internal structural term called a subsystem, which is a subset of the system that itself performs one or more specific functions. For example, each automobile system can be divided into several subsystems. The propulsion system is made up of a number of subsystems such as those that perform fuel delivery, lubrication, and cooling functions. The point is that the same structure of boundaries used to define systems can also be used to define "sub" or internal systems—each of the subsystems is also a collection of components working together to perform a given function.

The notion of a *super-system* is a natural extension of the aforementioned structure. Just as there are internal groups of system elements that can be defined as subsystems, there can be groups of systems that together perform even higher level functions. A super-system is defined as a collection of interdependent systems that together perform one or more functions.

To bring this discussion down to earth, let's develop possible super-systems using the human body and automobile examples. Let's start with the human body. Each of the body's systems by itself cannot define the entity. The function provided by the interactions of the body's systems is the delicate balance of function that we call life. On a much simpler level each automobile is a combination of systems working together to provide the function of transportation.

These examples illustrate how the concept can be applied to the functional hierarchy in nature and mechanics. The super-system structure is defined here for a reason that is more than formality. Our view of the world, including business functions and industrial processes is changing. Measurements can now be performed on a scale that was previously unimaginable. The super-system concept is a tool that enables us to step back from the traditional measurement convention of examining and quantifying functions of individual systems. Now, by using technology and our creativity, we can measure behavior on a more macroscopic level. The super-system definition allows us to recognize and examine the behavior of collections of systems that in some way depend on each other to provide an overall function.

Application of Measurements

Once measures have been properly defined and data have been collected, they must be translated to provide useful information about the systems and processes that were measured. This translation process is a potential mine field that must be carefully navigated to prevent gross errors in the use of the measurements. Their summary can result in a beautiful presentation that is completely misleading. As discussed in

Chapter 3, facts and figures, colored bar graphs, and pie charts can give the illusion of accuracy. The audience receives the impression that the presenter has searched through the muddy swamp of numbers and mined pure gold from the dark caverns of data bases, returning with all of the secrets heretofore hidden in the data. There is nothing wrong with this. In fact, every presenter of statistical information hopes their audience receives this impression! The problem arises because statistics are often used by someone who does not fully understand and properly employ them. Given a PC, a spreadsheet or "state" package, and a person armed with a little knowledge, the stage is set for an inappropriate application of statistics. Countless errors in statistical procedures can occur, ranging from mistakes in defining or interpreting sample bias, to using incorrect underlying assumptions, to misunderstanding of the general limitations of statistical parameters. Even something as seemingly simple and trivial as mistakes in the number of significant digits can completely distort statistical results. In short, the potential for both accidental and deliberate abuse is high. Not only are there many mistakes to be made, but to make matters worse, today's computer software makes the incorrect use and interpretation of statistics very *easy* to do as well. The real problem is that the results can look great but be completely wrong! Great *looking* results can convince their creator and the audience of an incorrect hypothesis.

The field of statistics in general has taken, and probably will continue to take, a lot of blame for supposedly making false statements. Just consider all of the controversy about the results of statistical polls around political elections. The *real* problem is not with statistics but with using or interpreting statistical results in ways that were not intended, thereby *mis*using them. Even when the statistical results are done correctly, however, it is important to remember that statistics can neither prove nor disprove. They can only supply information for people in the decision-making process.

The Mean Time Between Failure (MTBF)

In my experience, the most commonly misused statistic in industry is the mean time between failures. Exactly what is the *mean* time between failures? Suppose we want to calculate the MTBF over a given time period, beginning at time T_0 and ending at time T. During the time period from T_0 to T, there were n failures. For notation, let T_i, i = 1, 2, 3, . . ., n represent the successive times to system failure such that

$$T_0 < T_1 < T_2 < T_3 < ... < T_n < T \qquad (4\text{-}1)$$

The $\{T_i\}$ can be the dates of the year when the system experiences failure, number of machine hours, number of cycles, or whatever other time variable makes sense for the equipment. An underlying assumption here is that the variable of time used accurately accounts for the changes in the equipment. If equipment operates continuously or operates basically the same amount every day, then calendar time is proportional to actual operating time. This is not always the case. For example, with aircraft the use of calendar time is not the best indicator. Here, the number of takeoffs and landings and number of flight hours are indicators of equipment degradation and make better choices for the time variable. This is the same situation as would occur with an automobile. Most of us would agree that the mileage and type of driving are better indicators of a car's wear than its age. The lesson to be learned here? With MTBF, use whatever "time" units make the best sense. The mean *time* between failures does not necessarily have to signify calendar time.

The times between failure data, $\{t_i\}$, are computed from the values depicted in Equation 4-1:

$$t_i = T_{i+1} - T_i, \; i = 1, 2, 3, ... \; n - 1 \qquad (4\text{-}2)$$

Notice that the initial n failure dates produce only $n - 1$ data values denoting the times between failures. The time interval from the beginning of the time period, T_0, to the first failure is not used, nor is the time from the last failure, T_n, to the end of the survey period, T. There are two reasons that require this approach. The first has its origin in theoretical statistics. The time from the beginning of the arbitrarily set interval to the first failure has a different distribution form from the times between successive failures [1]. The second reason has its foundation in the English language. The MTBF says that the mean value is computed for the times *between* failures. Neither T_0, the beginning of the measured time interval, nor T, the end of the interval, is a failure. Using them to compute the MTBF would be inconsistent with the very definition of the statistic.

It is true that people sometimes do use the entire interval in computing the MTBF. From a practical perspective this is acceptable. After all, the reason for computing reliability statistics in general is to quantify system failure behavior. If the analyst feels the value obtained from this action is the most representative number possible, that's fine. It is, however, *not* the mean time *between* failures. It is the mean time between failures *assum-*

ing that failures occurred at the beginning and at the end of the period. Without the additional description, the MTBF use is just plain wrong.

There is another less common and relatively minor misuse of the MTBF statistic. This has to do with the particular form used to compute its value. In standard statistics there are three types of means: arithmetic, geometric, and harmonic. Which *Mean*TBF is used when? Unfortunately, the answer is "All of the above." Different versions of the mean are used in different situations. Generally, the arithmetic form is used for reporting failure time means, but the mean can be computed from another version. To complicate matters, there is also a generalized mean that contains the three aforementioned versions as special cases [2]. How do you know which one is used or which one you should use? The answer is that you don't know for sure, so the expression must be given along with the MTBF results. The point is that the name MTBF can mean different mathematical forms.

Using the times between failures, t_i, given in Equation 4-2, the three standard representations for the mean are given. The most common form is the *arithmetic mean*. It has the most wide range applicability in that it can accommodate negative and zero data values:

$$MTBF_a = \frac{1}{n} \sum_{i=1}^{n} t_i \qquad (4\text{-}3)$$

The *geometric mean* has uses where none of the values are zero or negative. It has applications in acoustics, queuing theory, and environmental science, just to name a few:

$$MTBF_g = \sqrt[n]{t_1 * t_2 * t_3 * K * t_n} \qquad (4\text{-}4)$$

The *harmonic mean* is generally applied to situations that have only positive numbers. It cannot be used for data sets that contain zeros. This form has particular application to circuit electronics problems. For example, the equation fits the mathematical relationship of resistors in parallel:

$$MTBF_h = \frac{n}{\dfrac{1}{t_1} + \dfrac{1}{t_2} + \dfrac{1}{t_3} + K + \dfrac{1}{t_n}} \qquad (4\text{-}5)$$

Failure data normally are positive numbers, namely the times at which something failed. In these cases, mean estimates will be ranked by the inequality:

$$\text{MTBF}_h \leq \text{MTBF}_g \leq \text{MTBF}_a \qquad\qquad (4\text{-}6)$$

Now that we've had a quick look at each of these means, let's look at some comparisons of them.

The arithmetic mean works well in representing a distribution if the values are not spread out too much. The geometric version de-emphasizes these differences to some degree. Take, for example, the arithmetic mean of the data set $\{1, 100\}$. This is $(1 + 100)/2 = 50.5$. The geometric version is $\sqrt{1 * 100} = 10$. From an order of magnitude point of view, 10 is in the middle of 1 and 100. From a linear perspective, 50.5 is the middle.

The latter two types of means are related. The arithmetic mean of the data logarithms is equal to the logarithm of their geometric mean. Thus, the geometric mean is particularly applicable when data values are logarithms.

Okay, so much for this theory, does anyone really use anything other than the arithmetic mean? The United States Environmental Protection Agency Regulations (40 C.F.R. 133.1022 (1987)) specifies minimum levels of effluent quality that can be obtained from secondary treatment in water treatment programs. The regulation did, at one point, state maximum pollutant levels in terms of the geometric mean of pollutant concentration computed from samples [3].

To give an example of where the harmonic mean is applicable, consider the following problem:

Suppose we drive 50 miles at 30 mph and another 50 miles at 60 mph. What is the average rate of speed for the 100-mile trip?

Before we compute the harmonic mean, let's first see what the arithmetic and geometric version give for their estimates of the average rate of speed, R.

$$R_a = (30 + 60)/2 = 45 \text{ mph} \qquad R_g = \sqrt{30 \times 60} = 42.4 \text{ mph}$$

Do either of these values accurately represent the average rate of speed over the 100 miles? No. In general, the arithmetic and geometric means do not accurately describe rate averages for mph nor for similar units such as cents per pound or barrels per hour. To answer the average speed question by direct computation, we must first account for the time travels at each speed. At both speeds, the same distance was traveled. Using the equation distance(d) = rate(r) x time(t), we find that traveling at 30 mph for 50 miles takes 1 hour and 40 minutes (5/3 hr) and traveling at 60 mph for 50 miles takes 50 minutes (5/6 hr). Together the entire trip of 100 miles took (100 + 50) = 150 minutes (2.5 hr). Thus, the average rate of speed over the 100 miles is r = 100 miles/2.5 hr = 40 mph. Now that we know the answer, let's compute the average using the harmonic mean expression:

$$R_h = 2/(1/30 + 1/60) = 2/(3/60) = 40 \text{ mph}$$

The *generalized mean* contains all of the above forms as special cases depending upon the value of the parameter, ω

$$\text{Generalized MTBF} = \left(\frac{1}{n} \sum_{i=1}^{n} t_i^{\,\omega} \right)^{\frac{1}{\omega}} \qquad (4\text{-}7)$$

where $\underline{\omega}$ $\underline{\text{Mean Type}}$
-1 Harmonic
0 Geometric
1 Arithmetic

Table 4-1 shows how the numerical results of the arithmetic, geometric, and harmonic means compare. The initial failure dates are used first to compute the times between failures. The data is then used to compute each mean version.

As shown in Table 4-1, the differences among the means are relatively large. The actual degree of the differences will vary in practice. This example is given just to show you that there are different ways of computing mean values, and in particular, the MTBF.

Figure 4-1 plots the generalized mean Equation 4-7 vs. the parameter, ω, using the data in Table 4-1. The values of the three standard mean relationships are identified. As you can see, a continuum of mean results are produced as ω varies over its plotted range.

As ω goes to minus infinity off the left side of the graph in Figure 4-1, the mean value approaches the minimum value in the data set. As ω goes to plus infinity, the mean approaches the maximum value in the data set. The generalized mean is a statistical quantity that has application primar-

Table 4-1
Comparison of Different Versions of the Mean

Failure Dates	TBF(days)	
Jan 6		
Jan 21	15	
Jan 22	1	Arithmetic Mean = 28.8 days
Feb 3	12	
Apr 14	71	Geometric Mean = 16.9 days
May 17	33	
Jul 17	61	Harmonic Mean = 6.4 days
Jul 20	3	
Aug 20	31	
Sep 3	14	
Nov 2	60	
Nov 18	16	

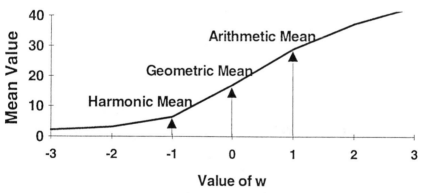

Figure 4-1. Generalized mean for table 4-1 data vs w.

ily in the theoretical aspects of statistics. It is presented here to illustrate that the mean can be different things to different people solving different problems.

There is another reason why a certain form for the mean is selected. It is also why the arithmetic version is the most common. The mean is a measure of the concept of "center." For example, the center of gravity of an object is the place where, if placed on a pin the object would balance. It is the location that can be used to describe the motion of the entire object. The formulas for this location are similar to the arithmetic mean. When measuring the reliability of a system, the MTBF is one way of measuring the "center" of the times between failure distribution. It is used in industry along with the standard deviation as a measure that represents the reliability of a system. This may or may not be true, as we will see in the next example.

Figure 4-2 shows four times between failure distributions. They all have the same mean and same standard deviation. The type of mean used in these calculations is the arithmetic form that has been theoretically determined to have the smallest error in a least square sense. The point of this discussion is that the MTBF may or may not be representative of the actual time between failure distribution. If these four distributions were times between failure results for a system or systems, the MTBF (even with the standard deviation computed) would hide the radically different behavior. In practice, it probably is not possible to develop actual time between failure distributions because limited data is usually accumulated.

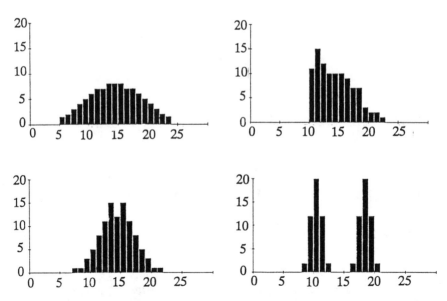

Figure 4-2. Time between failure distribution with identical MTBFs and standard deviations.

Just be aware that the underlying distribution is not given, suggested, or in any way implied by the MTBF and its standard deviation. The MTBF is a piece of reliability information, but it's not the whole picture. There is nothing wrong in using the MTBF, just make sure that it is being used for its designed purpose, and you know its utility *and* limitations. Be prudent with its application.

Availability

Availability is generally defined as the probability that the system will operate on a random demand [4]. In plain English, availability is the ratio of up time to total time. In mathematical terms, Availability (A_e), is expressed as

$$A_e = \frac{MTBF}{MTBF + MTTR} \qquad (4\text{-}8)$$

where MTBF = the mean time between failures

MTTR = the mean time to repair

If the formulas for MTBF and MTTR are substituted into Equation 4-8, the ratio becomes the time that the system was operating divided by the total time of the given period. This definition is also called *inherent availability* in some areas of application [5].

We use Equation 4-8 as a fundamental equipment-related availability metric because it describes the reliability of equipment regardless of its importance to system operation. This is admittedly an incomplete strategy because we are also interested in the operation and reliability of systems as they relate to production of product. Therefore, there are two other definitions of availability we use to measure the operational aspects: Production Time-Based Availability and Throughput Availability.

Production Time-Based Availability: A_t

$$A_t = \frac{MTBOF}{MTBOF + MDT} \qquad (4\text{-}9)$$

where MTBOF = mean time between operations failure

MDT = the mean operations down time

This definition is very close to Equation 4-8 except its reference is *system* operating time instead of *equipment* operating time. Because not all equipment failures affect system operations, this metric identifies events where the system has failed and caused production to stop. Systems fail because one or more equipment failures have occurred. The events that cause system failure are also events that contribute to equipment unavailability (1—availability), but the converse is not necessarily true. Equipment can fail (which changes A_e) with no loss of production (no change in A_t). For example, any system with on-line spares or other inherent engineered functional redundancies can have equipment failures without affecting production.

Production time-based availability, Equation 4-9, is designed to assess specific performance characteristics. Use it for measuring on-schedule production performance if your production is regulated to an optimized time schedule where any deviation from the program incurs penalties.

Throughput Availability

Throughput availability is the measure of availability that deals directly with production volume or throughput. Unlike the operations time-based version given in Equation 4-9, throughput availability adds a key variable, the time of the failure, to the measurement. This measure incorporates equipment failures that cause system failures (this equipment is often referred to as "process-critical equipment") and the *production penalty* of the failure from a simple throughput perspective. This version of availability has applications, for example, in situations where the product is directly used by the customer as it is produced, such as computer time available for transaction processing of reservations, financial activities, and customer orders. Throughput availability can be used as one metric in an overall strategy to supply information about the degree of customer satisfaction with service performance.

To calculate throughput availability, A_p, you first multiply the MTBOF by the production rate during this interval, resulting in the amount of throughput or product made during the time between system failures. Next, if the production rate at the time of failure is incorporated into the downtime calculations, the result is the lost production due to the system failure. Finally, you calculate A_p, defined as:

$$A_p = \frac{\text{Throughput}}{\text{Throughput} + \text{Lost Production Due to Failures}} \qquad (4\text{-}10)$$

Why would you want to use two measures of operations referenced availability? This question is best answered through this next example where the use of Throughput Availability has a very significant application.

Suppose your demand schedule for product is highly customer-driven, and it peaks every day around 12:00 noon. The production demand is shown by the bell-shaped curve in Figure 4-3. The two-system downtime

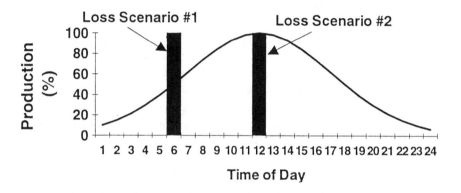

Figure 4-3. Down-time scenarios and production level.

scenarios are shown by the vertical bars indicating the system lost 100% of its production for one hour each at different times of the day.

The calculation of production time-based availability (Equation 4-9) yields the fraction, 23/24 or 0.958 regardless of the location of the one hour down time. It is the same for both scenarios. The production throughput availability definition (Equation 4-10) is applied to each scenario separately. The availability result of a failure during off-peak production time (scenario #1) is 0.960, while the same one-hour failure at peak (scenario #2) produces a throughput availability value of 0.920.

To continue the example, suppose production scheduling is designed to minimize the cost of electric power usage by operating at high production rates during low cost of power time intervals. Availability calculations should include a correlation between electric power costs and time-based availability. The reasoning is that as availability decreases, more production must be shifted to high electric rate times because the optimum schedule cannot be followed. The higher the availability, the lower the electric costs (to some limit). The converse is also true.

Even a simple concept like availability can take on many forms. All of these availability metrics, Equations 4-8--4-10, can be computed from data routinely compiled in most production reporting and computerized maintenance management systems.

There are two additional facts that affect availability: the time interval of reporting and the timing of failures inside the time interval.

First, let's consider the *time interval* over which the availability is computed. There are no specific guidelines on how long the interval should be, and clearly it is a function of the corporate need and culture. The intervals could be daily, weekly, monthly, etc. Just remember to explicitly include the time interval in the title of any availability report, e.g., "Production-Based Availability: January 1992." This seems like a simple, trivial point, however it is essential that the measurement be accurately described to the reader. From an accuracy in communications perspective, clarity of measurement information is not a revelation. From a mathematical perspective, the length of time used to compute availability can affect the result. This is why the time interval should always be noted. This fact sets the stage for discussion item #2.

Failure timing describes the effect of when the failure occurs within the time interval upon cumulative results. Let's suppose you are measuring availability on an annual basis, and reporting year-to-date results on a monthly basis. A failure at the beginning of the year will make availability *appear* lower than if the same failure had occurred later in the year. Here's how:

Consider the following availability calculations, both of which measure the availability reduction caused by the same 10-hour downtime event. We use production time-based availability in the example, but any version is suitable. All values are in hours. We'll compare the availability if the failure occurred in January with the same failure occurring in October.

January Calculation

For this example, suppose that operating time is 100 hours and down time to date was 4 hours.
Availability before a new 10-hour downtime:

$$A = \frac{100}{100 + 4} = 0.961 \qquad (4\text{-}11)$$

Availability after the 10-hour downtime:

$$A = \frac{100}{100 + 4 + 10} = 0.877 \tag{4-12}$$

Availability reduction due to 10-hour downtime = $0.961 - 0.877 = 0.084$

October Calculation

Now let's perform the same calculation in October when operating time is 2,100 hours and downtime to date was 80 hours.
Availability before the 10-hour downtime:

$$A = \frac{2,100}{2,100 + 80} = 0.963 \tag{4-13}$$

Availability after the 10-hour downtime:

$$A = \frac{2,100}{2,100 + 80 + 10} = 0.958 \tag{4-14}$$

Availability reduction due to 10 hour downtime = $0.963 - 0.958 = 0.005$

What does this mean? Simply that metrics, such as availability, become more and more insensitive to changes as data increases. This behavior should be recognized and understood when interpreting this type of statistical result. Consider the failure timing, when making judgments about availability results, and possibly when scheduling planned downtime events.

Availability of "Super-Systems" Vs. Systems

It is unusual today for a system to operate completely alone. Most often, systems are interdependent. Super-systems provide a framework to describe behavior on a larger scale. While availability is conceptually a straightforward metric, it can yield non-intuitive results when applied to super-systems. Super-systems are generally defined as macroscopic entities and are therefore directly connected to the bottom line of the business. The non-intuitive nature of super-system availability means that individual system availability may be high, but in effect, the overall availability can be much lower when perceived by customers and/or management. This is another reason to carefully define and communicate what is meant by the *system* when measurement parameters are defined. You can make the availability numbers look better or worse depending on how you define the system or super-system.

To explain this seemingly odd result, consider an example on the reduced scale of a motor and pump combination as a super-system. The motor and pump separately form the constituent systems. In this case, the systems are functionally dependent. The super-system fails if either the motor or pump fails. In the application of the super-systems concept to areas where the constituent systems do not have this degree of functional dependence, a failure of one system places the super-system in a failed "state." This mode of operation means the super-system can still be operating, but it is functioning at reduced efficiency, reduced effectiveness, or reduced safety.

Because the mean time between failures is essential to the availability calculation, we'll show how it can be affected by combining system data. Suppose the motor and pump each have a mean time between failures (MTBF) of 30 days. When both pieces of equipment are combined, the

Table 4-2
System Failure Data

System #1 Failure Date	Downtime (hrs)	System #2 Failure Date	Downtime (hrs)	System #3 Failure Date	Downtime (hrs)
17-Jan-92	9	17-Jan-92	6	11-Feb-92	8
8-Feb-92	80	29-Jan-92	109	18-Mar-92	8
22-Feb-92	12	8-Feb-92	8	4-Apr-92	8
7-May-92	20	8-Mar-92	54	11-Jun-92	85
9-May-92	4	2-Apr-92	3	6-Nov-92	24
10-May-92	10	5-Apr-92	63	16-Dec-92	8
23-May-92	2	9-Apr-92	8		
18-Jul-92	12	18-Apr-92	2		
1-Aug-92	4	10-May-92	8		
27-Aug-92	7	28-Jun-92	21		
8-Sep-92	4	1-Jul-92	4		
11-Sep-92	2	29-Jul-92	2		
7-Oct-92	39	13-Aug-92	13		
18-Oct-92	8	21-Aug-92	5		
9-Dec-92	18	27-Aug-92	36		
20-Dec-92	2	26-Sep-92	2		
7-Oct-92	8				
27-Dec-92	2				

super-system MTBF could be surprisingly different. Some possibilities are demonstrated using two systems' failure data over a two month period. This will be demonstrated by applying Equation 4-8 to the failure histories given in Table 4-2. This table contains failure histories for three systems. We define the super-system as the combination of them. The smallest time interval used in the analyses is one day. If two failures occur at the same time, only one super-system failure is observed. The explanation is the second failure is irrelevant because the super-system is already in a failed condition.

Now let's compare typical super-system availability with the availability results of each internal system. Table 4-3 gives the availability results for each system and the overall super-system. Notice that the availability of the super-system is much lower than the availability values for each of the constituent systems. In general, the availability of super-systems is not readily apparent but can be easily computed as a function of the constituent system data values. The super-system availability value must be

Table 4-3
System & Super-System Availability (Equation 4-8)

System #	Availability
1	0.972
2	0.959
3	0.981
Super-system	0.922

calculated directly. It is not simply a sum or an average of the availability values of the systems.

The practical importance of this mathematical behavior is that availability results can be highly dependent on how system failure data sets are and are not combined. In the case shown in Table 4-3, the "average availability" for systems 1, 2, and 3 is much larger than the super-system availability value:

$$\text{Average System Availability} = \frac{(0.972 + 0.959 + 0.981)}{3}$$
$$= 0.971 > 0.922$$

MTBF of Super-Systems vs. MTBF of Systems

Suppose system #1 failed on the first and last days of the first month and system #2 failed at the middle of both months. When the failure dates are placed in chronological order in the super-system, the resultant data points are three time between-failure values of 15 days each. As shown in Table 4-4, the super-system mean value is also 15 days.

Table 4-4
MTBF (Super-Sys.) < MTBF (Sys #1)
MTBF (Super-Sys.) < MTBF (Sys #2)

	System Failure Data	
Sys #1	**Sys #2**	**Super-Sys**
1-Jul-91	15-Jul-91	1-Jul-91
31-Jul-91	15-Aug-91	15-Jul-91
		31-Jul-91
		15-Aug-91
MTBF = 30	MTBF = 30	MTBF = 15

Another extreme situation is when the system failures occur on the same day. On both July 1st and 31st, 1991, both systems failed as shown in Table 4-5. The first failure on that day placed the super-system into the failed state. The second failure, while being of importance at the equipment level, was inconsequential at the super-system level. Because the motor failed and the pump failed during the same time interval, only one failure would be recorded for the super-system. The resultant time between failures is also 1 month.

Table 4-5
MTBF (Super-Sys.) = MTBF (Sys #1)
MTBF (Super-Sys.) = MTBF (Sys #2)

	System Failure Data	
Sys #1	**Sys #2**	**Super-Sys**
1-Jul-91	1-Jul-91	1-Jul-91
31-Jul-91	31-Jul-91	31-Jul-91
MTBF = 30	MTBF = 30	MTBF = 30

The same super-system result can be achieved from a much different failure behavior as shown in Table 4-6. In this case, the two systems' failures occurred as far apart as possible in the two-month period, under the constraint that both systems still have a MTBF of 30 days. In this case, the two systems shared one failure date, which is only counted once in the failure date recording for the super-system.

Table 4-6
MTBF (Super-Sys.) = MTBF (Sys #1)
MTBF (Super-Sys.) = MTBF (Sys #2)

	System Failure Data	
Sys #1	Sys #2	Super-Sys
1-Jul-91	31-Jul-91	1-Jul-91
31-Jul-91	31-Aug-91	31-Jul-91
		31-Aug-91
MTBF = 30	MTBF = 30	MTBF = 30

To present a situation where the MTBF for the super-system can be larger than either system alone, the system 30-day MTBF requirement is withdrawn. In this case each system has failed once in the last two months, exactly 1 day apart. The resulting MTBF in each case is 1 day. Combining systems and placing the failure dates in chronological order, the super-system MTBF is 20 days. The results are given in Table 4-7.

Table 4-7
MTBF (Super-Sys.) > MTBF (Sys #1)
MTBF (Super-Sys.) > MTBF (Sys #2)

	System Failure Data	
Sys #1	Sys #2	Super-Sys
1-Jul-91	30-Aug-91	1-Jul-91
2-Jul-91	31-Aug-91	2-Jul-91
		30-Aug-91
		31-Aug-91
MTBF = 1	MTBF = 1	MTBF = 20

There are numerous other examples that could be developed. These four cases are presented to demonstrate that the MTBF and availability of systems are not simply or intuitively related to the super-system MTBF and super-system availability. There are cases where intuition does work, but you cannot rely on it here. Super-systems may be more, less, or as reliable as the constituent systems. The only way to know for sure is to perform the calculations. The results are often very enlightening.

Total Productive Maintenance Measures

Classically, operations and maintenance personnel were separated by a philosophy of "I run it . . . you fix it." This philosophy nurtured differences in people's behaviors that caused a general lack of understanding, cooperation, and respect. The good news is that the wall between maintenance and operations is coming down. In many companies it no longer exists. One of the major reasons for this has been the introduction of Total Productive Maintenance (TPM). Starting in Japan in 1970, vast improvements in production, efficiency, and maintenance costs were realized by making some simple, basic changes in plant operations. The concepts of TPM are [6-7]:

1. Operators perform certain maintenance tasks on their equipment, e.g., cleaning, lubrication, etc.
2. Maintenance personnel perform operations-oriented functions.
3. Operations personnel assist mechanics in equipment repair.
4. Maintenance personnel assist in operation activities, e.g. equipment start-up and shutdown.
5. Inter-operation group interaction is required.

If there is one phrase to describe TPM it is "communication—understanding—cooperation." TPM requires no high tech investment. It is basically a bottom-up approach with management commitment to use people more effectively inside the plant. It empowers maintenance and operations personnel to solve problems.

There are several measures that people have applied to monitor TPM program improvements. The next section describes a good overall performance indicator termed "Overall Equipment Effectiveness," which combines several intermediate metrics classically used by themselves. This component of TPM produces a single measure with some interesting results.

Overall Equipment Effectiveness (OEE)

The three common measurements, availability, efficiency, and quality, are factors that together calculate Overall Equipment Effectiveness (OEE). In equation-like form:

OEE = Availability $\times$ Efficiency $\times$ Quality

The overall equipment effectiveness measure was defined in the discrete process industries, which must change equipment to run different products. In these industries, availability is defined as available time minus planned downtime and setup time. In other words, availability indicates how much time there is to produce product. Efficiency accounts for unplanned downtime such as from failures, jams and rate losses. Quality indicates spoilage, out of spec, and rework. Figure 4-4 indicates how these factors work together to compute OEE. The product quality indicator can only be applied to the effective time (or product) that is produced from the performance of availability and efficiency. The OEE measure has widespread applications in other industries.

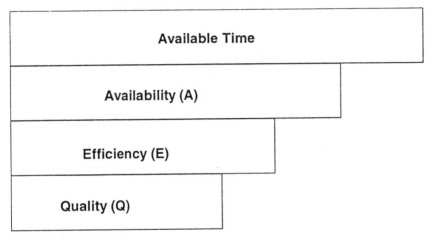

Figure 4-4. Visual representation of OEE parameter dependence.

Let's perform a simple, but interesting calculation of the OEE. Suppose availability, efficiency, and quality each have a value of 0.9. As far as each category is concerned, the production line is doing very well. Now let's calculate the OEE and look at this situation from an overall point of view.

OEE = 0.9 × 0.9 × 0.9 = 0.73 or 73%.

As this example shows, there is considerable room for improvement. If you compile the amount of potential savings or increased revenues from the lost 27%, there is a powerful motivation for management to promote and support a TPM effort. In industry, a typical OEE before TPM is 50% to 60%. It is estimated that after five years of TPM, the OEE should reach

85% [9]. This corresponds to an availability, efficiency, and quality of approximately 95% each.

With this indicator we can clearly see that although each individual measure appears acceptable, the overall operation is functioning at a rate which is significantly lower than any one measure indicates.

Computerized Maintenance Management Systems (CMMS)

Figure 4-5 shows the information flow of an efficient CMMS. Every activity on the right side of the figure indicates a "win" in the sense that action is being taken in a scheduled, planned fashion. The left side loop signifies the unwanted surprises; unscheduled failures that causes downtime, repair costs, and/or lost production. This is a "living" program in the sense that as more knowledge is obtained from the PdM (predictive maintenance) and PM (preventive maintenance) tasks, the frequencies and task contents can be changed to more accurately provide successful equipment operation. For example, as labor hours and materials are tracked in the course of performing scheduled maintenance, the labor time requirements and materials inventory can be refined to allow more accurate and efficient use of resources.

Initially, PM and PdM tasks and frequencies are loaded into the CMMS. As time advances, the CMMS automatically produces work

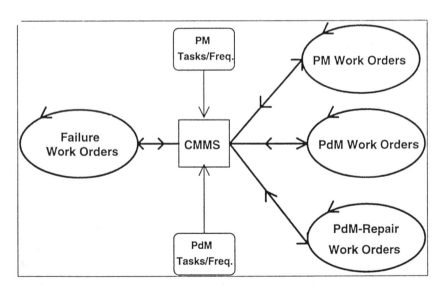

Figure 4-5. CMMS information flow.

orders for the scheduled PM and PdM events. These work orders detail the craft required and the specific items to be performed on the identified equipment. These activities are shown by the top and middle loops on the right side of Figure 4-5. The arrows on the loops indicate that the work orders are completed and the results recorded in a timely fashion. The CMMS manager or maintenance clerk closes the work orders by completing the appropriate fields in the work order computer screen layout. The information is then stored for future reference.

The lower right activity allows for additional work orders to be created from the initial PM-PdM work. This is done when a condition is identified in the PM-PdM session that warrants additional work to correct a deteriorating condition. Such proactive intervention prevents the condition from evolving into a failure event. The loop on the left side shows the documentation of work orders required to repair failures that do occur. The objective of this maintenance strategy is to have as much maintenance as practical on the right side of the work order flow and minimize the work orders issued for where the PM-PdM design has failed, such as unscheduled equipment failures.

In addition, there may be some equipment that is intentionally run to failure. These failures are included in the "Failure" work order category but do not signify a deficiency in maintenance policy. If certain equipment is selected to run to failure, the actual time of the event is unplanned so it qualifies for the "Failure" work order category. However, because it was *intended* not to perform preemptive maintenance on the equipment, the event is not a fault in the maintenance design. In essence, the failure was planned and therefore is not a weakness in the maintenance strategy. The event shares some properties of both the planned and unplanned categories, so depending upon how you think of these type of problems, you could put them really either place. My preference is to place such "run to failure" events in the planned category.

This type of event brings up an important point regarding the level of detail that is appropriate for different equipment. To allow users of the CMMS to track the key pieces of equipment properly, it is useful to stratify the level of data detail kept on equipment. Because data collection, entry, and analysis all take time, it makes sense to spend most of your resources with the most important systems or pieces of equipment. It doesn't make sense to keep the same level of detail for work orders required to cut the grass as for work orders required to maintain a major process-critical piece of equipment. Table 4-9 shows one stratification scenario where data detail levels are divided into four classifications.

Table 4-9
Equipment Data Record Keeping Classifications

Classification	Description	Data Detail Level
1	Process-critical equipment	Figure 4-5
2	Non-critical equipment	Figure 4-5 minus PM-PdM Repair
3	Run to failure equipment	Figure 4-5 using only PM & Failure
4	General work	PM work orders only

A conclusion that could be drawn from examining today's CMMSs is that they are data archival and work order generation systems. This is correct in almost all cases. The power of these seemingly simple functions is formidable. A CMMS provides a systematic, automated procedure for standardizing maintenance inside a plant [9]. The work orders it produces indicate exactly what tasks are to be accomplished every time they are issued. Work is scheduled in a set, periodic fashion. The savings in operational and maintenance costs generally have more than paid for the CMMS costs [10-11].

There are some issues associated with computerized maintenance systems that warrant mentioning. The three concerns I discuss here are the most subtle and common that I have seen in industry. They may seem obvious. However, I have observed several very successful companies that could be even more profitable if they only had realized the tremendous loss of productivity of their employees who are "hamstrung" and frustrated by working around a poor computer systems design.

The first issue is CMMS compatibility with other computer systems. Does the CMMS integrate easily with other pre-existing computer systems such as accounting, financial, and operations? Few if any people would argue against inter-system communication transparency. In practice, however, transparent inter-system functionality is seldom seen. There are numerous reasons for this fact. One is due to the computer technology itself. It is improving more rapidly than business can manage system changes. Companies have large amounts of time, money, and employee training invested in their technology. Upgrading computer hardware and software is a major corporate effort. The best strategy a company or corporation can develop includes a flexible, accepted, standard database software platform for data management in the different areas of the business. With this data structure uniformity, inter-system communication can be constructed using the company-wide standards.

The second issue is communications. The situation I am discussing here is one where the CMMS is accessible at several sites. The information systems could be centralized at corporate headquarters located in another state, or the CMMS could be limited to one plant with terminals connected by some type of network. The CMMS must be easily accessible to people who need the information contained within its databases so that they can change maintenance and operational practices to reflect the changes in equipment performance. This process is basic information feedback. If people perceive communications to the central databases as difficult, slow, or not beneficial, they are likely to develop and use their own private databases with the information they need to do their jobs. With the proliferation of personal computers having large data storage capabilities and fast processing speeds, workers can use low cost software to keep their own databases on their desktop machines. Keeping databases on personal computers may be fine as long as management recognizes the activity, and it is done in harmony with corporate information system policy. Without corporate direction is this area, there will be a general loss of efficiency in information management.

The last issue is best identified by the question "Who should choose CMMS software?" My response to this is somewhat generic: "The people who are going to use the tools should lead the selection of the tools." You might think that these people do not know the important technical details of database structure, hardware and software standards, and electronic communications to be best qualified to make the selection decisions. These "important technical details" will not make any difference if the software is not used. If the primary users of the software are not an integral part of decision process, their response can best be diplomatically identified as "non-ownership" resulting in little or ineffective use. The main point to remember is that a piece of software, regardless of its price and features, never improved a plant's reliability. This can be only accomplished by people. The software is an implement or tool to be used in combination with people's talents and experience. If these people are turned off with the CMMS product, what is gained? Now, of course the technical details are important for the reasons mentioned earlier. In selecting software, the lead role should be with the primary users, assisted and supported (but not led by) I/S technical team members. This is the best way to obtain the balance required to ensure that the software will be effectively used. A technically imperfect "software solution" effectively used is better than a technically perfect version that is ineffectively used. Remember, only people can change the reliability of the plant. There never has been a piece of software written that by itself changed the reliability of any system.

Summary

Measurement parameters and their applications provide indicators allowing us to understand, control, and manage industrial systems, including maintenance. Our challenge is to apply the best measurement parameters to the important aspects of the processes. System functions and their economic, safety, and operational importance supplies information on which parameters to apply and how to interpret their results.

References

1. Cox, D. R. and Lewis, P. A., *"The Statistical Analysis of Series of Events,"* Methuen, London, 1966, p. 36.
2. Casella, G. and Berger, R. L., "Deriving Generalized Means as Least Squares and Maximum Likelihood Estimates," *The American Statistician,* 46, no. 4, Nov 1992, pp. 279–282.
3. Finkelstein, M. O. and Levin, B., *Statistics for Lawyers,* Springer-Verlag, New York, 1990, pp. 29–30.
4. Lowery, E. E., Consolidated Lecture Notes: Tutorial Sessions: Reliability, Maintainability, and Statistics, ARMS, 1987 p. BM17.
5. Ibid., p. BM18.
6. Landwehr, W. R.,"TPM For Competitive Survival," Total Productive Maintenance Conference, The Manufacturing Institute, IIR July 1992.
7. Nakjima, S., *TPM Development Program,* Productivity Press, Cambridge, Mass., 1989.
8. Hall, R. K., "TPM:Total Productive Maintenance," Total Productive Maintenance Conference, The Manufacturing Institute, IIR July 1992, p. 2.
9. Williams, M., "CMMS Manages Buildings and Grounds, Infrastructure," *Maintenance Technology,* November 1990, pp. 65–69.
10. Valenti, M., "Maintenance Software Keeps Machines Up and Running," *Mechanical Engineering,* November 1991, pp. 63–65.
11. "CMMS Speeds Planning at Marion Steel," *Maintenance Technology,* January 1991, pp. 61–62.

CHAPTER 5

Reliability-Centered Maintenance: Description

These guidelines (RCM) provided the first formalized breakthrough in establishing new criteria for maintenance programs. They replaced maintenance concepts that had been in use for almost 60 years.

Lockheed Aircraft Official referring to MSG-1, the precursor of
RCM, 1976

The keystone of RCM is to develop maintenance tasks and frequencies for performing them, using the functional redundancies designed into many systems. What does functional redundancy mean? Consider the example of a pump system outlined below. The function of this system is to pump a product while complying with specific flow parameters such as suction and discharge pressures and flow rates. The system consists of primary and backup motor-pump units of equal capacity. The backup unit is activated by a pressure sensing device on the primary discharge side. The system is shown in Figure 5-1.

The system in Figure 5-1 is designed to function even if the primary motor and/or the primary pump fails, giving it functional redundancy. In this system, equipment failure does not imply system failure.

Using traditional maintenance methods, maintenance would be performed on both motors and both pumps exactly the same way, without accounting for the differences of their functions within the system. Maintenance work would be assigned for all equipment without considering whether or not the equipment was the backup or primary unit or, for that matter, if the system had a backup unit at all! In other words, traditional maintenance was designed with little or no consideration for how each piece of equipment contributed to the overall process.

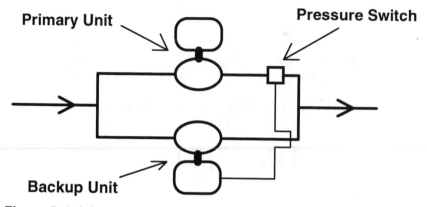

Figure 5-1. Inherent system redundancy.

Is there anything really wrong with this? Not entirely. However, in today's competitive fight for improved margins, cost reduction, and continuous improvement, the challenge to all plant employees is to operate and maintain industrial facilities in a safer, more reliable, and more productive manner with fewer resources. When a maintenance staff is challenged to increase availability and reduce costs with decreasing labor resources, the question is Where, What, and When should maintenance be performed? With RCM methods, maintenance designs use the advantages inherent in the system design. After all, maintenance departments are, and always have been, very interested in maintaining system functions. The shift in emphasis, however, is from the maintenance of *equipment operation* to the RCM theme of maintenance of *system function.*

With RCM, maintenance tasks and task frequencies are developed recognizing how each piece of equipment contributes to maintaining a system's function. Looking at our pump example again, this means that the backup unit will most likely have a different maintenance schedule than the primary unit. It may have the same tasks performed at different frequencies, or may have an entirely different set of tasks. The point is that the equipment's importance to maintaining system function is used in the development of a maintenance design.

The preceding discussion alluded to decision-making processes in RCM. RCM provides a mechanism for people to decide what maintenance needs to be done and also what maintenance does NOT need to be done. The essential common denominator of all versions of RCM is that engineering judgment and experience are vital to making decisions. The workers most familiar with the systems are key members in any RCM project team. There is no equation or automated way of translating the information into maintenance tasks and task frequencies.

The RCM method has many variants. Each has been developed from the method's basic principles. The method is not a cure-all or a magic bullet to resolve industrial problems. The success of each application depends on the support of management, the creativity of the RCM team, and the degree of cooperation of plant personnel involved with the systems.

These issues have surfaced as essential ingredients for successful RCM applications, regardless of the industry.

History of Failure

Since a main objective of RCM as a proactive maintenance strategy is to prevent failures, it is appropriate to discuss in detail here what is meant by the term *failure*. In the beginning of Chapter 2, the subject of failure was mentioned with regard to our perception of random events. As suitable measures are set in place, the seemingly random behavior is replaced with knowledge of system performance. Degradation of system operation can be monitored, and certain events that were previously categorized as failures can be avoided by scheduled intervention to repair worn components. Thus, the concept of failure changes as measurement applications change.

The dynamic nature of failure is reflected in its definition:

Failure ≡ the achievement of a performance characteristic or specified condition requiring unexpected or unscheduled maintenance

Let's examine the first part of this definition: "the achievement of a performance characteristic or specified condition." Equipment and components begin to age from the time they are built. One reason why the perception of failure continues to change is due to the changing nature of measurement applications. The measurements can be either procedural or technological based. Consider for example, failures of steam boilers [1]. A steam boiler explosion is clearly a failure. It was a failure 126 years ago and still is today. From the point of view of an explosion, nothing has and ever will change.

In Hartford, Connecticut, Thursday, March 2, 1854 was an unseasonably warm, sunny day. People were stretching their lunch hours to enjoy the early signs of spring. A mechanic at the Fales and Gray Car Works stepped into the engine room to participate in a conversation whose subject we will never know. While he was engaged in discussion a short distance from the engineer's post, the four-week-old boiler exploded with a tremendous force, completely destroying the boiler room and an adjacent blacksmith shop, killing 21 people, seriously injuring more than 50 others, and badly damaging the main building. The failure was attributed to "an excessive accumulation of steam . . ." that was allowed to occur due to the inattention of the engineer. The accumulation of excessive steam was clearly a performance condition that required unexpected maintenance.

Based on this experience, several procedural measurements were put in place to avoid any recurrence of this type of failures. Boiler operators became subject to a series of regulations designed to ensure their attentiveness. State boiler inspections were instituted, new boiler location sites were mandated to be away from population clusters, and engineering designs were explored that would not allow a boiler to produce more steam than it was rated to contain safely. All of these actions provide measurements. Their purpose is to increase our knowledge of the boiler's operational performance characteristics. The effect of these and many more measurement and design improvements has been the expansion of our definition of boiler failures. Today, performance characteristics and conditions requiring unexpected maintenance, still termed failures, are usually relatively minor repairs. This is a marked contrast to the original definition of failure as "explosion." The difference is that the repair is performed at a level that gives boilers a much higher margin of reliability and safety. Technology has given engineers and maintenance personnel very detailed measurement tools to observe performance characteristics and conditions. The threat of catastrophic events always exists; however, measurement tools provide the means to reduce *consequences* of failure by catching adverse, potentially harmful characteristics and conditions early, *before* catastrophic conditions can develop. The definition of failure is directly related to the sophistication of the facility. The more detailed the level of system measurement, the more detailed the criteria for failure.

Equipment failure behavior and the nature of failure are not simple, straightforward, or uniform. These subjects are continually discussed in the literature [2] and much has yet to be learned regarding the changing nature of failure. The definition of failure given here is subjectively accurate. It is virtually impossible to be quantitatively precise in defining failure, because the term or condition of "failure" means different things to different people, depending on the measures they have established. Even today, depending upon the industry, plant, level of technology application, and labor resources, the definition of failure can vary widely. The level of sophistication of a plant's operation influences its activities, philosophy, organization, measurement factors, systems integration, accounting, and overall company mission [3]. It also influences the specific criteria that define failure.

Traditionally, failures were events when something stopped working, that is, when it broke. Today, if a plant has an active and successful program for applying diagnostic evaluation technologies, a failure can be an emergency work order. In these cases, performance characteristics or conditions are observed and monitored. Wearout, accumulation of hazardous materials or gases, and other deterioration conditions can be detected. The "unexpected, but planned" maintenance event triggers an emergency work order to correct the condition.

Are failures going to be designed, regulated, and managed out of our systems? My answer is no. I believe that the number of "failures" will increase, but their consequences will be steadily reduced. The success of a maintenance strategy will not be assessed primarily based upon *number* of failures, but in concert with the *consequences* of the failures. In industry today, as the nature of failure changes, the effects or consequences of failures have been steadily reduced. To predict the not-too-distant future, the term "failure" will soon include corrective maintenance events with no production consequences.

Maintenance Classifications

There are two types of common classifications of maintenance tasks: *scheduled* and *unscheduled*. Here are the primary characteristics of each.

- Scheduled Maintenance Tasks: Planned work designed to maintain function on a regular, periodic basis. All maintenance designed to prevent failures falls into this category.
- Unscheduled Maintenance Tasks: Unexpected or unanticipated repair of equipment and components that have failed.

As far as structure is concerned, scheduled maintenance has the most ordered format. The goal of any maintenance organization is to *control*, though not necessarily minimize, the work that falls into the unscheduled category.

Scheduled Maintenance Tasks

Time Based

Tasks whose frequencies are determined from a known relationship between time (# of cycles, usage, age) with reliability (survival probability) *or* when conditioned-directed tasks are neither cost-beneficial nor technically practical. In other words, if the probability of failure is known to increase with time, then this type of task is appropriate. Remember that "time" is not necessarily calendar time, but should be defined as whatever "time" variable makes sense for your application. Here are some examples:

Corrosive Deterioration of Chemical Reactor Vessels

Most chemical reactor vessels are constantly in use. Experience with these vessels indicates that vessel thickness will be reduced below acceptable levels at fairly predictable intervals. If wall thickness readings can be obtained by non-invasive methods, then thickness as a function of time plots and various statistical methods can be applied to determine when turnarounds need to be scheduled. Such turnarounds are time-based maintenance.

Seal Replacements

Seal technology is a growing area where there still is plenty of room for improvement. It is nearly impossible to predict when these components are going to fail. Often, deterioration in other components causes seals to wear out at an accelerated rate. The observable equipment failure is in the seals, but the real failure may be caused by a condition not readily apparent to the observer or operator. It may not be cost-effective or even necessary to figure out what actually caused the seal failure. Simply replacing the seals on a regular basis as dictated by failure experience will solve the operational problem.

Conveyor Belts

Conveyor belts, rollers, and their components wear out over time. In these cases, calendar time may not be the "time" variable, but instead, the number of revolutions or cycles. For cars, tires are rated for a certain number of miles. The time variable here is miles or number of revolutions. For instance, 40,000 miles corresponds to about 36 million revolutions of a tire on a car. Conveyor belts and related equipment are similar to tire wearout in the sense that the number of revolutions is the wearout time variable.

Non-Process Motor/Pumps

This decision is purely an economic one. Sometimes it is more cost-effective to allow important non-process motors to run to failure rather than include them in a comprehensive maintenance program. In this situation, plants may choose to run small motors to failure and replace them with new equipment rather than perform overhauls when the motors begin to exhibit signs of wear. Motor size, maintenance department size, and motor importance to the process are all factors that together help decide what policy is best. An alternative to this program can be the assignment of maintenance tasks at specified time intervals.

Condition-Based

Condition-based tasks are those whose frequency is governed by the achievement of certain routinely measured conditions. Conditions warranting action occur slowly enough to schedule preemptive maintenance.

For example, in rotating equipment, high vibration readings coupled with high metal concentrations in oil samples and abnormal temperature indications may indicate bearing wear. These conditions would require preemptive maintenance to correct the symptoms *before* the equipment actually fails.

Failure Finding

Failure-finding tasks are used in situations where failures are hidden or not evident to the operators, such as with off-line or standby equipment. In this case, a task is designed that can readily discover potential failures. It is important to keep the failure finding tasks as simple as possible. For example, it is easier and faster to find a failure in a lighting circuit by testing the light than by measuring conductivity and resistance throughout the circuitry! Even though the failures these tasks may uncover are often not readily apparent in their effect on production, their importance must not be minimized. Because failure-finding tasks are applied to standby, safety, and other emergency systems that are activated infrequently, you will want to be sure that these systems will function in the event of an emergency.

Secondary Maintenance

Scheduled maintenance contains two major well known classifications: Preventive and Predictive. Preventive maintenance (PM) has become a term to denote all tasks that are time-based. Predictive maintenance (PdM) refers to condition-based tasks. Failure-finding tasks can conceivably fall into either area, but are usually time-based. There is another classification of maintenance tasks that is associated with the human reliability aspects of plant operation and maintenance. It is called Secondary Maintenance.

Secondary maintenance events involve human errors [4] and signify maintenance tasks that are required due to improper actions of plant personnel, improper operational procedures, or errors in design. In short, these events are maintenance-induced maintenance. It is easy to blame people for making mistakes. However, in almost all cases of secondary maintenance events, the person is not the root cause of the problem. Human reliability experts are now recognizing that managerial practices play a major role in causing secondary maintenance.

Here's an example where managerial procedures reduced reliability. In performing a statistical analysis of failures for a nuclear power plant, I discovered several situations where a component failed after a significant amount of operating time. A day or two later, after the problem was "repaired," it failed again. The description of work for the second failures had a common activity. The secondary maintenance involved activities like tightening loose connections, adjustments, and recalibrations on the equipment just repaired. In investigating the cause of these failures, I found that the competency of the maintenance personnel was *not* the root cause. Here's what I learned. The systems involved with the component failures were directly related to either power production or emergency standby and safety systems. In the case of power generation systems, any reduction directly affects profitability and performance, and is scrutinized by everyone from the stockholders to the Nuclear Regulatory Commission. As you can guess, the maintenance personnel are under a great deal of pressure to get the systems back on-line as fast as possible. Sound familiar? The stress of performing detailed, complex, and time consuming repairs under managerial and regulatory mandates can affect the quality of repairs, even with competent maintenance personnel. For standby systems in nuclear reactors, there is a 72-hour limit on repair time before regulations require the entire reactor be shut down, resulting in revenue losses of about $1 million per day. With this in mind, if maintenance said an emergency standby system repair would require a full 80 hours, there would definitely be an effort made to reduce the downtime to the 72-hour window mandated by regulation, and avoid a shutdown. This makes sense, and I am not saying that it is wrong to try to do this. But as you can see, managerial, and in this case regulatory, policies can place a tremendous amount of pressure on the maintenance personnel at the repair site.

Secondary maintenance errors are often difficult to identify. They are classified into three types and best described through examples.

Errors of Omission

- A mechanic forgets to check the oil level in the sight glass during routine inspection.
- An operator fails to open a valve before starting a pump.
- A pilot landed with the landing gear still retracted because he/she forgot to put the gear down.

Errors of Commission

- A maintenance person put the wrong lubrication oil in a gear box.
- An operator selected the wrong valve to open or close.
- After landing with the landing gear down, the pilot raises the gear instead of the flaps.

Cognitive Errors

- A maintenance person misreads a torque device and under-tightens bolts.
- An operator misreads a gauge or interprets the results incorrectly.
- A pilot lands with the gear up believing that he/she has lowered the gear.

Estimating the Proportion of Secondary Maintenance in Your Facility

How frequent are secondary maintenance tasks in today's industrial facilities? This area is of growing concern and a subject of current investigation. It has been estimated that 20–30% of all work orders written in a plant fall into this category [5]. One U.S. Air Force study found that 40% of all maintenance required to restore a sample of F-4 Phantom jets to operational condition fell into this category [6]. Human errors and their contribution to overall risk are discussed in Chapters 10 and 11, so we will defer a detailed discussion until then. Clearly, however, secondary maintenance tasks deserve more recognition. People do not know beforehand that they are going to forget something or perform a task incorrectly. No pilot I knew ever took off intending to land with the gear up!

Suppose we want to estimate the proportion of work orders (maintenance tasks) that fall into the "secondary" category. Because secondary maintenance tasks can be identified only after their execution, we can develop the proportion by looking at a set of work orders that are already completed. The work orders must be reviewed by a maintenance analysis team whose composition reflects both technical expertise and judgment integrity. The team members need to understand the technical aspects of the process and equipment maintenance. They also need an impartial view of the tasks to minimize, and ideally exclude, bias in the subjective decisions. To start, the team chooses a random sample of work orders and analyzes each one individually. They then judge whether or not the maintenance falls into the secondary maintenance category. It's a relatively simple process, and the results can be very enlightening.

The random sampling procedure is fundamental for the method discussed here. This signifies that work orders must be chosen without any identifiable pattern. In other words, you must choose work orders by a procedure that gives every work order an equal chance of being selected. How you do this is up to you. You can draw work orders out of a hat or use a random number generator to select them. It is acceptable to limit the set of potential work orders, such as to one year's worth of data. You need some manageable set of data, but do not take the easy way out by choosing a serial sequence of work orders, because this procedure would not give every work order an equal chance of being chosen and negates the validity of the sample size vs. accuracy procedure we'll discuss next.

The literature indicates that the proportion of maintenance classified as secondary maintenance generally ranges from .2 to .4. Let's initially use a value of .3. Also, we will accept an accuracy of ±5% with a 95% confidence level. The next question is "How many work orders need to be analyzed to get this accuracy in the result?" To determine the sample size of work orders, refer to Figure 5-2.

Following the arrows in Figure 5-2, notice that the sample size required for ±5% accuracy is roughly 320. This number is independent of the plant being analyzed, assuming that secondary maintenance accounts for .3 or 30% of the work orders. If the desired precision is increased to ±4%, the sample size is significantly increased to about 500. If you only desire accuracy of ±10%, then fewer than 100 work orders are necessary. This means that the computed estimate of the proportion would be 30% ±10%.

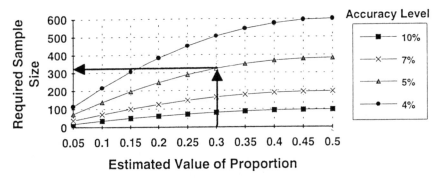

Estimated Value of Proportion

Figure 5-2. Sample size vs accuracy: proportion estimation at 95% confidence.

Because estimates for the proportion are not known before the study, how do you know the sample size initially? From an initial proportion estimate, the sample size can be approximated. Let's say that we choose the following set of parameters:

$$\text{initial proportion estimate} = 0.3\ (30\%)$$
$$\text{desired accuracy} = \pm 5\%$$

We will always use a confidence level of 95%.

Now, suppose from the roughly 320 work orders, the proportion of work orders in the secondary category turns out to be 40%. Figure 5-3 can be used in a reverse fashion to determine the resulting accuracy of the statistical experiment.

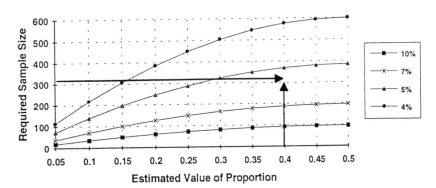

Figure 5-3. Sample size vs accuracy: proportion estimation at 95% confidence.

By interpolating between the curves, the actual 95% accuracy of the proportion can be determined. In this case, the accuracy is roughly ±6%. As you can see, unless the initial estimate is much different from the one computed from the sample, the initial sample size probably will be acceptable. Small differences in experiment accuracy (±4%, ±5%, or ±7%), regardless of the proportion value, make a big difference in required sample size. Depending on the time your analysis team has available, you can use the figures above to quantify the accuracy of secondary maintenance statistical experiments, regardless of the observed value of the proportion.

RCM Terminology

Because RCM focuses on the maintenance of system function rather than equipment operation, we now turn our attention to definitions required to develop the RCM method. The terms are defined in descending order of logical ownership.

System

An overall plant or plant sub-section that has been identified for RCM analysis, this term is discussed extensively in Chapter 4 along with its relationship to a super-system.

Subsystem

An assembly of equipment and/or components that together provide one or more functions and can be considered a functionally separate unit within the system. The unit has physical boundaries and quantifiable inputs (called "in interfaces") and outputs (called "out interfaces"). Mathematically speaking, the intersection of all subsystems is zero, and the union of all subsystems is the system. Chapter 4 also addresses this term in its relationship to the system and super-system.

Functional Failure

Each subsystem performs certain functions. Functional failures describe how failures of each function can occur. These descriptions generally do not denote equipment-specific problems, but describe functional problems such as a full or partial loss of power, erratic voltage readout, loss of fluid-air boundary (leaks), or high liquid level.

Failure Mode

A failure mode identifies each specific equipment-related condition that can cause the loss of the subsystem's functions. Failure modes describe how equipment failures must occur to cause functional failures. In this sense, it incorporates the inherent reliability of the system. For example, consider a pumping subsystem designed with an on-line spare activated by a pressure-sensing switch. For the functional failure "loss of pressure," one failure mode could be "loss of primary pump *and* pressure switch failure." Together these two component failures would cause the functional failure. For highly redundant systems, the identification of all the various ways the functional failures occur can take considerable time and effort. The more redundant and complex the subsystem, the more time and effort are required to identify the failure pathways. The key point is that the detail of failure analysis needs only to be at the level where maintenance tasks can intervene. Classically, this type of analysis has used fault and event tree methods to ensure that all failure pathways have been covered. For highly complex systems, Sneak Analysis may also be applied.

Failure modes may refer to the root cause, depending on the level of the analysis, but usually refer to the observed failure effect such as seal failure, motor failure coupled with pressure switch failure. Root cause implies that a scientific analysis can determine the fundamental beginning of the failure. For example, a pump failure effect may be observed as a cracked impeller. The root cause, then, is whatever caused the rotor to crack. We would always like to know the root cause of every failure. Most of the time, however, limited resources and time only permit this level of detail for the most serious failures. RCM can and usually is accomplished without detailed system fault and event tree analysis. In most cases, these powerful tools require too much time and labor resources. System schematics may also be used. Tool selection depends on the application and the objectives of the analysis. The key resource for developing the failure modes for the functional failures is the plant experience of the employees.

All of the functional components are interrelated. Each failure mode specifically pertains to one functional failure that is contained in a specific subsystem of the entire system. This "natural" interdependence will be used later to develop an efficient, shorthand way of keeping track of all the variables. From a visual perspective, Figure 5-4 shows how the functional components are related.

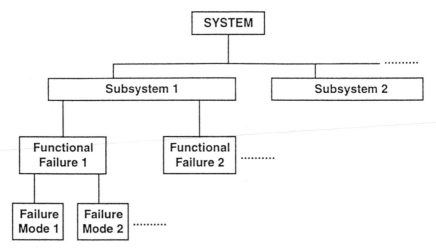

Figure 5-4. Interrelationships of RCM functional components.

The Five Steps of Reliability-Centered Maintenance

A major part of the RCM method is to take a large system, divide it into smaller, separate and simpler systems called subsystems, determine the functions of each subsystem, and determine what can cause each function to fail. This process is called system function decomposition and is described by steps 1 through 3. In step 4, the failures are categorized according to their criticality or importance. Finally, step 5 determines the actual maintenance tasks and schedules to mitigate the expression of the failures. These steps are detailed below.

1. Define System and Subsystem Boundaries

The system is divided into mutually exclusive subsystems with separate, non-overlapping boundaries. Everything that crosses these interfaces is identified. Each subsystem has "in interfaces," indicating what comes in to the subsystem, and "out interfaces," indicating what goes out of the subsystem. Everything going into and out of the system must be identified whether it is product, steam, electricity, control signals, or anything else. The artificial boundary helps ensure that all of the important equipment necessary for system function is included in the analysis.

Subsystems must not have boundaries that overlap. This would make interface definitions impossible. The interface concept is fundamental to the method. In physically small systems such as aircraft, the definition of subsystems is somewhat easier because the components are physically close to each other. For industrial applications, subsystems can physically cover many miles. For example, consider a cooling subsystem with heat exchangers and a cooling tower. This subsystem can weave its way in and out of a plant's other subsystems. Care must be exercised in defining subsystems in this large scale. Make sure that the subsystem functions are separate from other subsystems. This exercise may require looking at the plant from a new and different perspective. This part of the RCM project can take longer than expected. It is worth the time. If this step is not done properly, succeeding steps in the RCM method can become confused, and the result may be a loss of valuable information.

2. Define Subsystem Interfaces, Functions, and Functional Failures

For each subsystem, "in interface" quantities are transformed to the "out interface" quantities by the functions of the subsystem. These functions and how they can fail are enumerated in this step. Functional failures describe the different ways a subsystem can fail to perform its functions and do not necessarily identify specific equipment or components. The functional failure analysis, sometimes referred to as FFA, identifies the specific ways the output interface quantities cannot be produced and the functional failures that are internal to the subsystem.

3. Define Failure Modes for Each Functional Failure

Specific equipment and component failures that cause each functional failure are identified. It is at this level in the functional hierarchy where equipment and components are usually brought into the analysis. The dominant failure modes are developed from a failure modes and effects analysis (FMEA). The FMEA supplies cause/effect information and identifies the specific conditions that must be prevented by preemptive maintenance action. This is the most detailed level of the functional decomposition. It must be performed accurately and completely because it is from these identified characteristics that PM/PdM tasks will be determined.

4. Categorize Maintenance Tasks

For each failure mode, a series of yes/no questions are asked. Based on the sequence of answers, failure modes are matched with a type of task. Next, the consequence level is established for each. The set of questions is the same for all failure modes, regardless of their functional failure and subsystem locations. This procedure applies a uniform standard to task classification that helps ensure adequate treatment of all failure modes, regardless of who is performing the categorization. Consequence levels are labeled as "criticality classes" and are a function of which pathway is followed in the sequence of answers. The set of yes/no questions is called a decision tree. An example is shown in Figure 5-5 [7].

Notice the first question determines if the failure is observable. Hidden failures are found primarily in standby equipment designed to function as on-line backups to primary systems, or when certain operating conditions trigger its function, as with emergency pressure relief valves or emergency fire spray systems. Due to the nature of these systems, standard maintenance tasks are not appropriate because the systems only function when needed. Many times they are only needed in emergency conditions at which time standby system function is imperative for safe shutdown or continued operation. In the best of situations, they are never used in their entire useful life cycle. On the other hand, they only need to work once to save severe losses and justify their entire life cycle expense. Maintenance tasks for these systems are denoted as "failure finding." Failures are discovered basically by testing the system.

In Figure 5-5 the four letters in bold denote the "criticality class" that each failure mode is assigned in each of the categories. Category A is for failure modes that affect safety. Category B failure modes are not safety-related, but do affect operation. Category C modes do not affect safety or operations, but can potentially be prevented by maintenance tasks and thereby save costs. Category D tasks are modes suitable for failure finding tasks.

The decision tree logic provides a standard format for matching the ways subsystem functions can fail to maintenance tasks that can prevent them. In the situation where a category A failure mode cannot be matched with effective maintenance tasks, there are two possible choices; (1) accept the hazard or (2) perform an engineering redesign. The objective of the redesign is to remove the failure mode altogether or, at the very least, change its criticality classification.

The type of task and frequency of application are variables that are defined by the people involved with the study. There is no equation, special formula, or computer program for this process. The RCM construct provides a framework for deciding these things, but the final decisions are based on engineering experience and judgment.

The result of applying this logic is a correlation of failure modes matched with maintenance tasks to be implemented in the most efficient manner. There is another, equally important list. This is the set of failure

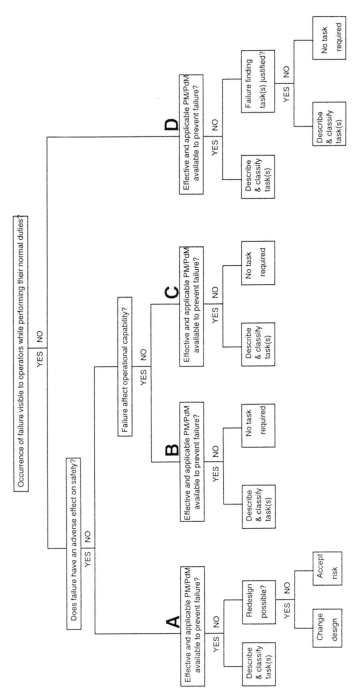

Figure 5-5. RCM decision tree example. (Copyright © 1986. Electric Power Research Institute. EPRI NP-4795. *Use of Reliability-Centered Maintenance for the McGuire Nuclear Station Feedwater System.* Reprinted with permission.)

modes that do NOT need maintenance. The power of the RCM methodology is that it provides an engineering, system-function justification for performing and NOT performing maintenance. As mentioned, excessive maintenance can actually decrease system reliability. Knowing where to and where NOT to allocate maintenance resources is essential to a successful maintenance design.

5. Implement Maintenance Task

By this point, maintenance task requirements have been defined. The last step of the RCM method is to group tasks and match them to available labor resources. This process indicates the size of the work force required and provides information about the skills required for scheduled maintenance. If current staff requirements cannot be effectively matched with the maintenance tasks, then the cost penalty can be estimated or a reevaluation of the maintenance tasks requirements can be performed. If the available staffing exceeds the maintenance task requirements, then the excess labor can be redeployed. If the number of available personnel is fewer than the number required by the maintenance tasks, then the penalty is the cost of repair and loss of production resulting from the expression of the failure modes for which no maintenance is assigned.

The purpose of this step is much more than a simple job assignment problem. It is a very powerful procedure that *measures* the effectiveness of the maintenance program. This type of measurement is not a formula computation, but a sophisticated, comprehensive process that considers the inherent system design, available maintenance technologies, and available labor resources in the context of the overall cost of operation.

Numbering System

There is a natural hierarchy generated by the functional dependence of a system to its subsystems, functional failures, and failure modes. Each failure mode is unique to a functional failure, each functional failure is unique to each subsystem, and each subsystem is unique to each system. I have developed and used an index system to simplify tracking of the large volume of highly structured data. This numbering system is useful in quickly identifying items as subsystems, functional failures, and failure modes and their relationships.

The index system is presented from the perspective of a super-system. Normally an RCM project will address only one system with many subsystems. Historically, RCM has been used for detailed system studies

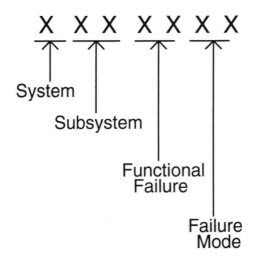

Figure 5-6. Graphical description of an RCM index system.

where one system was extensively analyzed. In the future, RCM will be used for more extensive maintenance design studies encompassing groups of systems in a streamlined fashion. The index code is developed for this generalized case. Its structure is versatile and allows for simplification without any logical change in numerical ordering.

The index hierarchy for the functional dependencies of a super-system is based on the use of six digits. The leftmost digit identifies the system. The next two digits identify the subsystem number. Digits 4 and 5 identify functional failures of the subsystem. The last two digits refer to failure modes of a functional failure. For example:

1000000: System 1
1020000: Subsystem #2 of system #1
1020300: Functional failure #3 of subsystem #2 of system #1
1020304: Failure mode #4 of functional failure #3 of subsystem #2 of system #1

Figure 5-6 shows the relationships between the numerical and functional hierarchies. Notice that subsystem *functions* are not explicitly considered by the numerical index. This is because subsystem functions are included in the enumeration of functional failures. From a practical perspective, the functions themselves are NOT the targets of maintenance. The maintenance design does not change how a subsystem functions. The main focus of reliability-centered maintenance is the maintenance of these functions, whatever they are. Thus, what *is* important is how the functions can fail, that is, the functional failures. Maintenance tasks are designed to guard against having these things happen. The main targets of preemptive maintenance actions are preventing the functional failures. This is why sub-system *functions* are not indexed.

The following is an example of a super-system numerical index for a gas compressor plant consisting of two compressor systems and a cooling tower. The super-system is taken as the entire plant. Each compressor system and the cooling system (including the cooling tower) have the rank of "system" in the numerical hierarchy:

System #:	1000000 Compressor #1
Subsys#	**Description**
1010000	Power Source
1020000	Gas Compression
1030000	Liquid/gas sep. (Knockout Drum #1)
1040000	Liquid/gas sep. (Knockout Drum #2)
1050000	Liquid/gas sep. (Knockout Drum #3)
1060000	Lube/Seal Oil
1070000	Steam Condensing

System #:	2000000 Compressor #2
Subsys#	**Description**
2010000	Power Source
2020000	Gas Compression
2030000	Liquid/gas sep. (Knockout Drum #4)
2040000	Liquid/gas sep. (Knockout Drum #5)
2050000	Liquid/gas sep. (Knockout Drum #6)
2060000	Lube Oil
2070000	Seal Oil
2080000	Steam Condensing

System #:	3000000 Cooling System
Subsys#	**Description**
3010000	Stream Cooling

References

1. Weaver, G. and McNulty, J. B., *An Evolving Concern Technology, Safety and The Hartford Steam Boiler Inspection and Insurance Company 1886–1991,* The Hartford Steam Boiler Inspection and Insurance Company, 1991, pp. 1–3.
2. Moubray, J., *Reliability-Centered Maintenance,* Industrial Press Inc., 1992, pp. 187–228.
3. Peterson, S. B., "From CMMS to RMS: The Reliability Management Lifecycle," *Maintenance Technology,* May 1993.
4. Ryan, T. G., Haney, L. N., Ostrom, L. T., "Crucial Role of Detailed Function, Task, Timeline, Link, and Human Vulnerability Analyses in HRA," ARMS, 1993, pp. 490–497.
5. Oliverson, R. J., private communication, March 1993.
6. Smith, C., Civ. OASD (MRA&L), "Why and What is Reliability-Centered Maintenance?" unpublished paper, May 12, 1976, p. 5.
7. Kessler, S. F., "The Reliability-Centered Maintenance Study at the Fast Flux Test Facility," January 1988, DE88-014649, p. 13.

CHAPTER 6

Reliability-Centered Maintenance Example: The Bicycle

Nothing is particularly hard, if you divide it into small jobs.

Henry Ford

Up to this point, we've discussed RCM in a largely theoretical manner. The basic principles of RCM have been applied in different ways by different people with different objectives, so its description varies from source to source. All of the applications are RCM, but each has special attributes that were required to fit individual project objectives and constraints. Descriptions of RCM can be difficult to translate into personal experience because the fundamental principle of *maintenance of system function replacing maintenance of equipment function* is more philosophy than substance. A literary scholar can write a book describing in eloquent and artistic detail how strawberries taste. However, unless the reader has tasted the fruit, the real essence of the taste cannot be understood through any use of words—some things must be experienced.

It's time to taste the strawberries. The RCM method has been described in words. Now let's walk through its application with a familiar system, the standard, one-speed coaster bicycle (Figure 6-1).

The five major steps in RCM, detailed in Chapter 5, are listed as follows for easy reference as we walk through this example:

1. Define system and subsystem boundaries
2. For each subsystem
   - Define subsystem interfaces
   - Define subsystem functions and functional failures
3. For each functional failure
   - Define failure modes

116

Figure 6-1. This chapter uses the example of a bicycle to show how to apply the RCM method.

4. Categorize maintenance tasks

5. Implement maintenance tasks

We will apply each of these steps in sequence and develop our RCM analysis results in a spreadsheet format. As we complete each step, the results of our analysis will be shown, as exhibits, surrounded by a border, distinguishing them from the supporting discussion and explanations. The subsystems are studied in order according to our numerical indexing system, described in Chapter 5. Let's begin with step 1.

Define the System and Subsystem Boundaries

The System

For this example, we'll define the system as the bicycle alone, without its operator or rider. There is no super-system application in this example. In this case, we will use a very simple system definition. In industrial situations, when a continuous process is often occurring, it is not always as easy to define boundaries of a system. Most of the time, system boundaries are logical, as process or product changes suggest system boundaries definitions.

The Subsystems

Process changes also help in selecting boundaries for subsystems. Each of the subsystems of the bicycle corresponds to the major processes that together compose the functions of the bicycle. They are identified and illustrated in Figure 6-2.

The subsystems are entered below in our spreadsheet, with index numbers assigned. There is no special logic used to assign index values to specific subsystems. For all general purposes, the assignments are arbitrary. The beginning of the RCM structure is given in Exhibit 6-1. The system level index is omitted because there is only one system being studied, and there is nothing gained by carrying the extra digit.

When partitioning a system into subsystems, the only firm guidelines are with subsystem boundary definitions. *Subsystem boundaries must not*

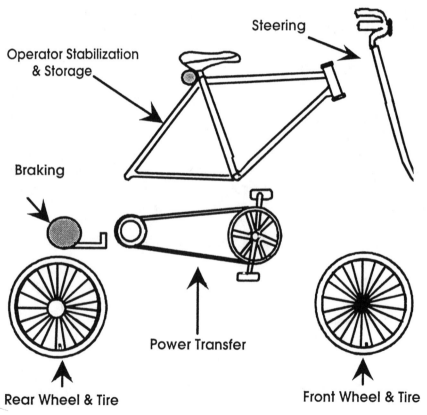

Figure 6-2. Bicycle subsystems.

Exhibit 6-1
Subsystem Assignments for Bicycle RCM Analysis

10000	**Power Transfer**
20000	**Braking**
30000	**Support and Stabilization & Storage**
40000	**Front Wheel & Tire**
50000	**Rear Wheel & Tire**
60000	**Steering**

overlap. Other than this constraint, you are free to define boundaries in any way that makes sense to you. If you start the analysis and find you want to divide or combine subsystems, corresponding boundary interfaces also need to be changed. This is usually not difficult because of the mutually exclusive (non-overlapping) boundary principle that is fundamental to RCM. The interface definition process actually helps you to decide if your boundary definitions are correct. If you can accurately identify all subsystem input and output quantities, that is, the in and out interfaces, chances are the boundary definitions are consistent and non-overlapping. This completes Step 1.

Next, Steps 2 and 3 will be applied together on each subsystem. Let's begin with the first subsystem.

Define Subsystem Interfaces, Functions, and Functional Failures

Subsystem 10000: Power Transfer

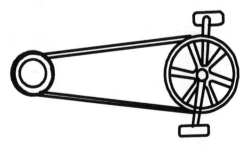

Description: Input forces are received through the pedals and transmitted to the **Rear Wheel & Tire** and **Braking** subsystems via the pedal support bracket, primary gear, and chain.

It's important to provide a brief description of each of the major components of the subsystems. The subsystem documentation contains items that are either required for the RCM analysis, are useful for future reference by people not involved with the analysis, or are included to refresh the memories of those who did the work. The documentation level used in this example illustrates the basic type of information needed for the RCM analysis. The type, amount, and level of written detail for any RCM project must be decided on a case-by-case basis. Keep in mind that *up to a point* the higher the level of written detail the more valuable the analysis. The results and methodologies can be used by people not involved with the project either at a later time or at a different facility with similar systems.

The **Power Transfer** subsystem is used to shift operator-supplied forces to the secondary gear located in the **Rear Wheel and Tire** subsystem. There are many force transmission or force transfer functions involved with a bicycle. These words will be used many times in the descriptions of subsystem functions. This is not surprising because the main purpose of a bicycle is to transfer operator supplied forces into system movement. The pedal surfaces are examples of some of the subsystem boundaries. The "input forces" supplied by its operator, the bicycle rider, pass through these boundaries and are transmitted through the structure to the rear gear as shown in Figure 6-3. Once the force arrives at this location, it is transmitted out of the subsystem.

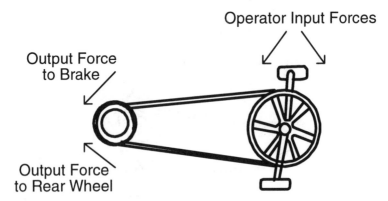

Figure 6-3. Force transmission diagram.

Define Subsystem Interfaces

Imagine yourself on the subsystem boundaries and observe what passes through to the inside. Drawing a picture that isolates the subsystems, such as Figure 6-3, may help in identifying all of the inputs. The following are the quantities that are observed going into the subsystem.

> **In Interfaces**
> Operator supplied forward pedal forces (acceleration and cruise)
> Operator supplied backward pedal forces (deceleration)
> Resistance force from **Braking**

Notice that not all of the inputs occur all the time. This is to be expected. Here we are identifying the specific items that can go into the system, not their frequency and duration. For example, bike riders input more forward pedal forces than backward forces because more time is spent accelerating and cruising than decelerating. These differences in the frequency and duration of forces are irrelevant in determining the interface quantities.

The last in interface is a response force from another subsystem. This is a natural product of the non-overlapping subsystem boundary requirement of RCM. What goes out of one subsystem can go into other subsystems.

Using the in interfaces and Figure 6-3, notice that as you depress (input force) one pedal, the other pedal experiences an equal but opposite (output) force. Also, output forces are directly related to input forces. However, depending on the direction of force, the output force is directed towards different subsystems.

> **Out Interfaces**
> Forward force to **Rear Wheel** and **Tire** (acceleration and cruise)
> Backward force to **Braking** (deceleration)
> Operator supplied pedal forces

Define Subsystem Functions and Functional Failures

Subsystem Functions

Now that we have a good picture of what goes in and out across the interfaces, we can focus our efforts on determining all of the functions performed by the subsystem.

A pinch of ingenuity must be used here. The scale of subsystem functions must be at the level of detail desired in the RCM analysis. *Also, the detail level in identifying functions must be at a level where maintenance is reasonable.* For example, a function of the **Power Transfer** subsystem could be to transfer force from each chain linkage to the next chain linkage. From a functional standpoint this is fine, but from a practical view no one is going to take the time to inspect every structure and pin of the chain linkage. The larger scale functions "Transfer forces to chain" and "Transfer forces from chain" contain the essence of the functions we are trying to describe at a level where practical maintenance can be applied.

This is one major area that can bog down an RCM project. People can get so tied up in the functional details that they lose sight of the overall maintenance design concept. Again, it is important to define functions at a level where maintenance can have an effect.

Here is a sample list of functions for the **Power Transfer** subsystem. Notice that each of the functions begins with a word showing the action of the function. The last function is a sophisticated way of saying that the bearings' and axles' seals should not leak. The language used is not intended be to esoteric, but accurate. Grease leaks or leaks of any kind are NOT functions. The proper statement of the function is to maintain grease/air boundary.

Functions
Transfer operator forces from one pedal to the other pedal
Transfer forward operator forces from pedals to chain
Transfer backward operator forces from pedals to chain
Maintain lubricant/air boundary

Functional Failures

Generally, there is at least one functional failure for each subsystem function. This is not an absolute rule, however. Sometimes functions can be omitted from a functional failure listing if the essence of the failure is covered under another. With the **Power Transfer** subsystem, we have an example of this situation. Also, some functions can have more than one functional failure. The functional failures listed in the following sections itemize how the subsystem functions can fail in a manner that is important at a level maintenance can be practically applied. Also keep in mind that at this stage in the analysis, we are still dealing primarily with functions and their failures, not specific equipment. Before we can identify equipment failures, we must know their importance or contribution to providing subsystem functions.

The identification of in and out interfaces is required to ensure that the subsystem boundaries are non-overlapping and to assist in the accurate determination of the functions. As mentioned earlier, the purpose of this procedure is to determine how subsystems can functionally fail so that the final maintenance design can prevent these failures. The enumeration of subsystem functions is an intermediate step in this process. We index only the functional failures because they are what need to be prevented through the application of specific maintenance and operational procedures yet to be determined.

The functional failures are listed using the index system discussed in Chapter 5. The RCM spreadsheet for the **Power Transfer** subsystem now has the form shown in Exhibit 6-2.

Exhibit 6-2
RCM Spreadsheet for Power Transfer at Functional Failure Level

10000	**Power Transfer**
10100	Loss of operator force transfer to pedals
10200	Loss of forward force transfer to chain
10300	Loss of backward force transfer to chain
10400	Failure of lubricant/air boundary

Define Failure Modes for Each Functional Failure

It is at this level in the analysis where equipment failures are considered. Let's take the first functional failure listed in the spreadsheet: **Loss of operator force transfer to pedals.** By studying the subsystem mechanisms, what equipment must fail for this functional failure to occur? This is the question that is asked for each functional failure. These failures or, for redundant systems, combinations of equipment failures, are the failure modes. For the first functional failure, the following failure modes are identified. Note that any one of these equipment-related failures will cause the functional failure to occur.

Functional Failure: Loss of operator force transfer to pedals
Failure Modes:

1. One pedal fails
2. One pedal bearing fails
3. Both pedals fail
4. Both pedal bearings fail
5. Poor pedal surface

The RCM spreadsheet is now updated with the failure modes determined for the subsystem functional failures. The failure modes are the most elemental information in the RCM functional hierarchy. To make reading the listing easier, the functional failures are shown in bold-face in Exhibit 6-3. This completes Steps 2 and 3 for the first subsystem.

Exhibit 6-3
RCM Spreadsheet Component for Power Transfer Subsystem

Index	Description
10100	**Loss of operator force transfer to pedals**
10101	1 pedal fails
10102	1 pedal bearing fails
10103	Both pedals fail
10104	Both pedal bearings fail
10105	Poor pedal surface
10200	**Loss of forward force to chain**
10201	Chain fails
10202	Primary gear fails
10203	Foreign object between chain & primary gear
10300	**Loss of backward force to chain**
10301	Chain fails
10302	Primary gear fails
10303	Foreign object between chain & primary gear
10400	**Lubricant/air boundary failure**
10401	Primary gear axle seal fails
10402	Pedal seal fails

Notice that the Functional Failures 10200 and 10300 have exactly the same failure modes. As far as the chain is concerned, there is no difference between forward and backward forces. Consequently, it is no surprise the failure modes are the same. There are two reasons for distinguishing between forward and backward forces here. First is that they effectively go to different subsystems. The forward force goes directly to the **Rear Wheel and Tire** subsystem, while the backward force goes into the **Braking** subsystem. Second, the "importance" of the forces are different. This specific point will be discussed later in the chapter.

Now let's move on to the second subsystem where we will apply the same procedures.

Subsystem 20000: Braking

Description: Input forces from the secondary gear located on the rear wheel to a high friction stationary surface inside the hub. The resistance force is transmitted back to the **Rear Wheel and Tire.**

Identify Subsystem Interfaces

The **Braking** subsystem plays almost a passive role in the functioning of a bicycle. Like braking systems on most vehicles, it must apply decelerating force when required and function satisfactorily by NOT applying the decelerating force in all other cases.

There are two active inputs to this subsystem. They are the forward and backward forces from the chain located in the **Power Transfer** subsystem. Each of these two forces is intended to produce a different subsystem response. This is why the forces are separately listed. The third input (and output) interface quantity is from the structural framework that holds the coaster brake lever in place. The input force is really a response force from the bike frame (**Support and Stabilization**) subsystem. Even though this force is a response from another subsystem, we list it to assure that all input quantities are identified.

In Interfaces
Forward force from **Power Transfer**
Backward force from **Power Transfer**
Coaster brake lever response force from **Support and Stabilization**

The output forces can be identified from the quantities that go into the subsystem. They are listed below. Notice that because there are no subsystem responses for forward force applications, no output force is listed. In this case, the number of in interfaces is not the same as the number of out interfaces. There is no reason to expect the number of respective interface quantities to be the same or, for that matter, directly related.

> **Out Interfaces**
> Deceleration force to rear wheel
> Coaster brake lever tension

Define Subsystem Functions and Functional Failures

Now we ask the question: What functions does the braking subsystem perform? Examine the differences between the in interfaces and the out interfaces. These changes are created by functions in the subsystem. In this case, the **Braking** subsystem provides the deceleration force to the subsystem that can assist in decelerating the bicycle, and it prevents the same force from being applied when not desired. Also, because there is a lubricant internal to the braking mechanism located within the rear wheel, this lubricant must remain in its intended area and not leak out. The three functions are now given as:

> **Functions**
> Provide deceleration force to **Rear Wheel and Tire** from backward force
> Prevent deceleration force to **Rear Wheel and Tire** from forward force
> Maintain lubricant/air boundary

There is one interesting point about this listing that illustrates the functional approach to maintenance inherent in RCM. Before you read further think about this. If you ask people in general what is the function of brakes on a bicycle, most people would say that the purpose of brakes is to stop the bicycle. Notice that stopping the bicycle is not a function of the braking subsystem. The purpose of the brakes is to apply a deceleration force to **Rear Wheel and Tire.** In other words, brakes stop wheels. The braking subsystem, along with the synchronous function of other subsystems, stops the bicycle. The same is true for other vehicles such as automobiles, airplanes, snowmobiles, and trains. An exception can be seen in the following example. Suppose you are sledding down a hill on a cardboard and you notice you are heading right for a large tree. You apply the brakes, your feet. In this case, the brakes do stop "the car," which in this case is you. Your feet may or may not stop the cardboard, but who cares!

Proceeding now with the braking subsystem analysis, how can the above functions fail? Well, we can lose the decelerating force. We can have the decelerating force applied when not desired, and/or the lubricant can leak out. These failures follow directly from the functions. However, the implications of losing the decelerating force depend on the *degree of loss* from both a functional and a maintenance perspective. Maintenance planning must separate "total loss" from "partial loss," because the mechanism that produces the decelerating force wears out over time and a degradation of performance is to be expected. Thus functional failures can reflect the degree of failure.

Exhibit 6-4 shows the functional failures for the **Braking** subsystem.

Exhibit 6-4
RCM Spreadsheet for Braking at Functional Failure Level

Functional Failures

20100	Partial loss of deceleration force
20200	Total loss of deceleration force
20300	Deceleration force applied to rear wheel from forward force
20400	Loss of lubricant/air boundary

Define Failure Modes for Each Functional Failure

Up to this point in the analysis, very few details of the coaster braking mechanism itself have been mentioned. The interfaces, functions, and functional failures were defined at a system rather than a component level. The details of the braking subsystem were not required. Now we need to balance the level of subsystem detail with the level of possible maintenance action. Usually, the result is closer to the functional (macroscopic) level than the component (microscopic) level. This is because maintenance actions are based upon observable quantities and not upon micro-level details of a system. People must respond to how a system performs, using its inherent design to ensure continued function. In this context, the details of the system design really do not matter.

For example, consider the operation and maintenance of a motor-pump unit that is constantly operating. Do people really care what the internal design is? At the time of purchase, these details may have been considered, but right now when this unit is pumping product, people are now more interested in keeping it running. The question becomes: What can be done to keep this unit from failing and to keep it operating at an acceptable performance efficiency?

The same is true in this bicycle example. The failure modes are developed basically from a system, or macroscopic, perspective *at the level of detail required to maintain the desired functions*. It is through these pathways that people will identify how to maintain an acceptable level of sub-system function operation. Too much detail can cause an RCM project to fall behind schedule because of the numerous technical failure modes that "could" happen. People can get so involved with the detailed failure modes that they lose sight of the overall objective. The dilemma that RCM project analysts constantly face is the balancing of current maintenance actions, available maintenance actions, possible maintenance actions, project costs, and labor resources with the level of failure mode detail. Although some of these details may be important, I have found that it is better to have too little than too much. If the results obtained from the "too little" approach indicate that more detailed analyses are warranted, then you have some justification for doing the additional work and extending the project duration, and perhaps project cost. The problem with developing overly detailed failure modes is that project progress can slow to a point that RCM may seem too overwhelming and expensive to continue.

Exhibit 6-5
RCM Spreadsheet Component for Braking Subsystem

Index	Description
20100	**Partial loss of deceleration force**
20101	Brake pad wearout
20102	Chain slipping
20200	**Total loss of deceleration force**
20201	Catastrophic brake pad failure
20202	Brake friction lever failure
20203	Brake gear failure
20204	Brake rotor failure
20205	Coaster brake lever bolt-nut removal
20206	Coaster brake lever failure
20300	**Deceleration to rear wheel from forward force**
20301	Brake locked
20302	Foreign object in brake assembly
20303	Chain twisted
20400	**Loss of lubricant/air boundary**
20401	Seal failure in brake assembly
20402	Bearing failure
20403	Covering puncture

Exhibit 6-5 contains the listing of failure modes for each identified functional failure. This list identifies samples of how the loss of functions may occur. Those of you who are very familiar with the coaster bicycle brake may be able to identify additional failure modes that could be mitigated by standard or new maintenance tasks.

Subsystem 30000: Operator Support and Stabilization

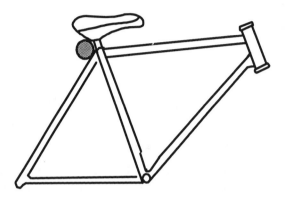

> **Description:** This structure stabilizes operator interface control and allows the operator to continue operation. The subsystem is also responsible for maintaining the stationary positions of the other subsystems.

Identify Subsystem Interfaces

This subsystem is the rigid, structural backbone of the bicycle. It is a major conduit for force transmission and for response forces from the other more visibly active subsystems. The braking mechanism itself is not included here because it is supported by the **Rear Wheel and Tire** subsystem. The coaster brake lever that is attached to the frame is mentioned, however.

Because this subsystem is primarily structural, forces going in must come out. Therefore, the out interfaces follow directly from the in interfaces. The subsystem distributes the operator weight along its structure to the connected subsystems. The subsystem is designed to play a passive role in that the resultant forces are intended to keep the connected subsystems aligned.

In Interfaces
 Operator weight
 Steering response forces
 Power Transfer response forces
 Rear Wheel and Tire response forces
 Front Wheel and Tire response forces
 Coaster brake lever response force

Out Interfaces
 Operator weight
 Forces required to keep connected subsystems in stationary positions

---

Define Subsystem Functions and Functional Failures

Function identification is important. Each must be separately identified because we are designing a maintenance plan to maintain these functions. This cannot be done unless we know what the individual functions are.

The subsystem functions are fairly simple to identify. For example, the operator must be able to sit comfortably on the seat without it wobbling. It must be secure and not easily moved. The structure itself must be able to support the weight of the operator and be adjustable for the leg length of the rider.

Functions
 Support operator weight
 Provide variable seat position on adjustment
 Maintain **Steering** support and position
 Maintain connected stationary positions of all connected subsystems

---

The next function, "Maintain **Steering** support and position," illustrates how subtle subsystem boundaries can be. The outside frame of the steering column is part of the **Support and Stabilization** subsystem as its function is to maintain support and position for the **Steering** subsystem. The steering column itself and the separator rings and various other components, however, are assigned to the **Steering** subsystem because they are actively involved with the steering function. Although the components of these two subsystems are physically integrated, they remain separate subsystems based on their functions. The same principle is applied when identifying the functions involved with maintaining the position of the other subsystems.

How can these functions fail? The operator weight function is easy. Either the subsystem supports the weight or it does not. There is no middle ground. The same is true for the seat position. For the steering related function, two separate functional failures are identified, indicating the degree of failure that is possible. Notice that the failure of the steering support and position function essentially negates the entire **Steering** subsystem functions. The remaining functions are passive in the sense that when they are performed, nothing moves. The function may seem trivial, but without it, the bicycle system would not function. In fact, it would fall apart. The last functional failure combines all of the subsystem support functions into one listing. These results are shown Exhibit 6-6.

Exhibit 6-6
RCM Spreadsheet For Support and Stabilization at Functional Failure Level

Functional Failures	
30100	Unable to support and stabilize operator
30200	Unable to provide seat adjustment
30300	Partial loss of **Steering** support
30400	Total loss of **Steering** support
30500	Unable to maintain subsystem positions

Looking at the list of functional failures in Exhibit 6-6, one thought that comes to mind is that the severity or importance of these functional failures to the operation of the overall system has not been considered. For example, the total loss of steering is clearly more serious than a loss of seat adjustment. I bring up this point now to show the large contrast in functional failures' importance illustrated by this subsystem. The entire topic of failure importance is what Step 4 of the RCM method is all about. We will postpone further discussion of this point until then.

Define Failure Modes for Each Functional Failure

Elaboration of the particular ways loss of function can occur is provided by the descriptions of the failure modes. Each functional failure is now considered individually, and the mechanistic equipment or component level pathways that can cause the functional failure to occur are identified. Exhibit 6-7 lists the results of this analysis.

Exhibit 6-7
RCM Spreadsheet Component for Support
and Stabilization Subsystem

Index	Description
30100	**Unable to support and stabilize operator**
30101	Catastrophic structural failure
30102	Seat structural failure
30103	Seat covering failure
30200	**Unable to provide seat adjustment**
30201	Tilting adjusting nut missing or rusted
30202	Tilting adjusting nut threads stripped
30203	Vertical adjusting nut missing or rusted
30204	Vertical adjusting nut threads stripped
30205	Catastrophic seat failure
30300	**Partial loss of Steering support**
30301	Minor structural failure
30302	Steering support column cracked or bent
30400	**Total loss of Steering support**
30401	Catastrophic structural failure
30402	Catastrophic steering support column failure
30500	**Unable to maintain subsystem positions**
30501	**Rear Wheel and Tire** support failure
30502	**Front Wheel and Tire** support failure
30503	**Power Transfer** support failure
30504	Catastrophic structural failure
30505	Coaster brake lever support failure

There is another way to analyze this subsystem, starting with determining the subsystem's functions. **Support and Stabilization** has an important function that affects all other subsystems: keeping the bicycle together. The structure keeps the rear wheel, front wheel, and the pedals in their respective places. To describe these functions through the complete identification of the connected subsystems would involve specifying functions like:

- Maintain **Rear Wheel and Tire** support and position
- Maintain **Front Wheel and Tire** support and position
- Maintain **Power Transfer** support and position
- Maintain position of coaster brake lever

There is nothing wrong with doing this, but it is not likely that the increase in detail will add any accuracy. If these functions are identified separately as indicated, the functional failures lead to only one failure mode. Generally, any time you obtain a functional failure with only one failure mode, you should consider a re-organization that assigns more than one failure mode to every functional failure. This saves analysis time and also helps ensure the functional description is correct. I am not implying that a functional failure with only one failure mode is absolutely wrong. Just be aware that in some cases, re-organization *may* provide a more accurate and efficient point of view.

As a side discussion to the bicycle problem, one type of functional failure that may have one failure mode is "loss of product/air boundary." Here the failure mode could be just "seal leaks." On the other hand, if the product is under high pressure, you may want to consider having two failure modes relating to the magnitude of the failure. In this case, separating the large and small size of the leaks recognizes that each has different safety, operational, and maintenance implications.

Subsystem 40000: Steering

Description: Steering contains the handle bars and the shaft that passes through the frame, ending at the slots where it is fastened to the front wheel axle. The frame is a part of **Support and Stabilization** while the inside core, the part that turns, belongs to **Steering.** The vertical and level adjustment components of the handle bar also belong to this subsystem.

Identify Subsystem Interfaces

The interfaces quantities are as follows:

In Interfaces
    Position maintenance forces **Support and Stabilization**
    Operator weight forces from **Support and Stabilization**
    Control forces from operator
    Response forces from **Front Wheel and Tire**

Out Interfaces
    Operator weight forces to **Front Wheel and Tire**
    Control forces to **Front Wheel and Tire**

Support and Stabilization holds this subsystem in place. This is the intent of the wording of the first in interface. Note that the out interfaces arise directly from the in interfaces. They represent the transmission of forces through the subsystem. The in interface, "Response forces from **Front Wheel and Tire**," identifies both the forces that come from the wheel's resistance to directional change due to its rotation or gyroscopic force, and forces originating from the tire's effect on the system boundary, namely the ground.

Define Subsystem Functions and Functional Failures

The functions of this subsystem follow directly from studying how the in interfaces are transformed or transmitted to out interfaces.

Functions
    Transfer steering controls to **Front Wheel and Tire**
    Transfer operator weight forces to **Front Wheel and Tire**
    Maintain stationary position of **Front Wheel and Tire**
    Assist in operator position security

Here is an example where the functional failures correlate directly with the functions. The only exception is that the steering control is broken into two items to model the degree of steering loss. More attention will be given to this topic in the discussion of the next subsystem.

Why is the steering column separated from the wheel? What is the benefit of dividing them into separate subsystems? The reason I chose to do this is related to the function of "Maintaining stationary position of the **Front Wheel and Tire.**" The steering structure has a notch on each side to support the front wheel axle. This is an important function because if one lip of a notch shears off, the front wheel could literally fall off the bicycle, or at the very least, become twisted and abruptly terminate forward motion. By placing the supporting structure and the front wheel in different subsystems, it is easy to identify a "function" of **Steering** as maintaining the position of the **Front Wheel and Tire.**

Let's consider what would happen if we combined these two subsystems into one. The function "Maintain stationary position of **Front Wheel and Tire**" becomes an internal event not represented by examination of in and out interfaces. If we were not cognizant of this potential failure, we could possibly overlook it in analyzing the combined subsystem. The combined subsystem is not incorrect, but it raises the potential for overlooking functional and failure modes that exist. Thus, when you are initially partitioning any system into subsystems, keep in mind the larger the subsystem, the larger the potential for missing important internal functions that can negatively affect maintenance design. With this bit of advice, the **Steering** functional failures are given in Exhibit 6-8.

Exhibit 6-8
RCM Spreadsheet for Steering at Functional Failure Level

Functional Failures
40100   Partial loss of steering control
40200   Total loss of steering control
40300   Unable to assist in operator position
40400   Unable to accept operator weight forces

Define Failure Modes for Each Functional Failure

The handle bars are assumed to have two adjustments. One bolt and nut combination is used to fix the bar's height and the other to change their angled position in front of the rider. A partial loss of steering control can occur if these adjustments become loose. Also, foreign matter between **Steering** and the frame will reduce steering capability. Total loss of control will occur if major, catastrophic adjustment or structure failure occur

Exhibit 6-9
RCM Spreadsheet Component for Steering

Index	Description
40100	**Partial loss of steering control**
40101	Handle bar vertical position adjustment nut loose
40102	Handle bar rotational position adjustment nut loose
40103	Foreign object between **Steering and Operator Stab.**
40200	**Total loss of steering control**
40201	Foreign object between **Steering and Operator Stab.**
40202	Steering support failure
40203	Catastrophic handle bar failure
40204	Catastrophic vertical adjustment failure
40205	Catastrophic rotational adjustment failure
40206	Catastrophic vertical bar failure
40207	Front wheel position bracket failure
40300	**Unable to assist in operator position**
40301	Catastrophic handle bar failure
40302	Catastrophic vertical adjustment failure
40303	Catastrophic rotational adjustment failure
40304	Catastrophic vertical bar failure
40400	**Unable to accept operator weight forces**
40401	Steering support failure
40402	Front wheel position bracket failure

among the identified subsystem components. The failure modes for the other functional failures are self-explanatory.

Listing failure modes, as we do again in Exhibit 6-9, should not be taken lightly. The team must develop a balance between too few and too many failure modes. That is, a balance between detail and practicality is required for successful completion of an RCM project. If, during the development of your failure modes lists, you identify additions that are *significant* to the maintenance plan development, great! However, sooner or later you will be constructing more and more unlikely, possibly imaginative scenarios. Eventually, you will ask yourself the question every RCM analyst asks: How much is enough? When do you state that the failure mode list for all practical (and perhaps even for some impractical) purposes is sufficient? There is no easy answer. The more systematic the method for identifying the failure modes, the less serious the concern. A procedure using the system's design, historical data, and human-based operating experience can do the best job if managed properly. The point here is that functional failure and failure mode identification can be done successfully at a reasonable cost, and completed in a reasonable amount of time using system knowledge, available data, and the experience of the personnel familiar with the process under study.

Subsystem 50000: Front Wheel and Tire

Description: The front wheel accepts directional controls from the **Steering** subsystem and maintains forward motion by rotation with the inflated tire providing suspension and force damping functions. The spokes are attached to the hub and wheel rim, keeping the circumference of the wheel and tire at a uniform distance. The hub is lubricated by two bearings located on each side. The hub itself attaches to the **Steering** subsystem by two nuts also located on each side of the hub.

Identify Subsystem Interfaces

The interfaces for this subsystem are developed again by placing an imaginary wall around the subsystem and identifying what passes across the boundary.

In Interfaces
    Steering control forces
    External (from system boundary) tire forces
    Operator weight forces from **Support and Stabilization**

Out Interfaces
Gyroscopic forces from rotating wheel
Response force to **Steering**
Operator weight forces from **Support and Stabilization**

Forces going into this subsystem are from **Steering** control, the external surface, and from the weight of the operator. Output forces are the gyroscopic force from the wheel rotation and forces responding to the movement of the **Steering** subsystems. These forces act directly on the axle at the location where **Steering** attaches to **Front Wheel and Tire.** Last, the operator weight is transmitted through this subsystem to the external boundary.

The front and rear wheels look the same, and they do share some common functions. There are, however, significant differences that warrant treating these areas as separate subsystems. The interfaces we've just seen show that the **Front Wheel and Tire** subsystem has connections to the external system boundary, **Support and Stabilization,** and to **Steering.** The functions of the rear wheel subsystem have a different interface environment than that of the front wheel. Comparison of the interfaces and functions of the front and rear wheel subsystems clearly shows that just because areas have similar equipment, they do not automatically belong to the same functional subsystem. In general, equipment similarities have little or no relationship to subsystem formation. What determines the equipment and component grouping into RCM subsystems is the functional relationships of the hardware—not their type, manufacturer, or physical makeup.

Define Subsystem Functions and Functional Failures

The functions are determined by considering how the in interfaces are transformed into out interfaces and by studying what functions are operating solely inside the subsystem.

Notice, in Exhibit 6-10, that specific areas of the subsystem have been identified to assist in understanding the procedure used to identify subsystem functions.

The functional failures follow the functions fairly closely with two exceptions. First, the general function of wheels, "Transfer rotation into linear motion," is not explicitly mentioned in the functional failures. This function is implicitly included in the other more detailed functional failures.

Functions

Transfer **Steering** control forces to wheel (hub and wheel connections required to turn wheel)

Transfer rotation into linear motion (general function of wheels)

Provide external force damping to wheel (tire)

Maintain lubricant/air boundary (seals)

Maintain wheel roundness (spokes and rim)

Exhibit 6-10
RCM Spreadsheet for Front Wheel and Tire at
Functional Failure Level

Functional Failures

50100	Partial loss of rotation capability
50200	Total loss of rotation capability
50300	Total loss of external force damping to wheel
50400	Loss of lubricant/air boundary
50500	Loss of wheel roundness

Second, loss of external force damping to wheel is divided into "partial" and "total" categories. This is because the subsystem can gradually lose this function as well as fail instantly. To account for this fact, two functional failures are used—one which indicates continued operation at a reduced performance level, and one indicating complete failure. Partial versus complete failure can also be addressed in the identification of the failure modes. An example of this will be mentioned when we reach Step 3 for the subsystem.

In practice, people may be willing to accept reduced performance on a limited basis. The maintenance design being developed by the RCM analysis needs to reflect operating states that may occur in the subsystem's normal and abnormal conditions. The degree to which abnormal conditions are addressed is strictly a judgment call. The fundamental principle is that the RCM-derived maintenance design is intended to maintain system function. How subsystem functions can fail, even partially, is an essential part of the analysis. Most of the time, subsystems will fail in the partial mode more than they will by catastrophic events. Thus, it makes good sense to design a plan that reflects the relatively high frequency, low consequence failures as well as the low frequency, high consequence failures. Again, this part of the RCM analysis does not consider the *importance* of these functional failures. This will come later. Right now the objective is to describe the practical and possible ways the subsystem functions, and how it can functionally fail.

Define Failure Modes for Each Functional Failure

We again ask the question, "What can cause each functional failure?" or "What equipment, components, or combination of components must fail to cause functional failure?" Let's consider the failure modes by functional failure as shown in Exhibit 6-11.

Let's look at the loss of rotational capability, 50100 and 50200. Bearing failure and an object between the steering structure and the wheel can clearly cause the indicated functional failures. The failure severity signi-

Exhibit 6-11
RCM Spreadsheet Component for Front Wheel and Tire

Index	Description
50100	**Partial loss of rotation capability**
50101	Bad hub bearings
50202	Foreign object between **Steering** and wheel
50200	**Total loss of rotation capability**
50201	Major hub bearing failure
50202	Foreign object between **Steering** & wheel
50203	Hub bolt failure
50204	Major spoke failure
50300	**Loss of external force damping to wheel**
50301	Slow tire leak
50302	Fast tire leak
50303	Rim dented
50400	**Loss of lubricant/air boundary**
50401	Hub-axle seal failure
50402	Hub-axle bearing failure
50403	Hub-axle covering puncture
50500	**Loss of wheel roundness**
50501	Dented rim or wheel
50502	Major spoke failure

fies whether it is a "partial" or "total" event. Loosing one or both hub bolts or a major failure of many spokes will also cause a total loss of function.

Now consider 50300, "Loss of external force damping to wheel." In plain English, this functional failure means the tire goes flat. You can have a fast or slow leak in the tire, or the rim can be bent to such a degree that the tire cannot hold the seal. We are assuming that the tires are tubeless, because if the tire had a tube, the loss of tire-rim seal would not necessarily imply the tire would go flat. Even with tubeless tires, it may be possible for the rim to be bent and maintain tire pressure. The operator might, however, feel the cyclical thumping of the dented rim.

Functional failure 50400, "Loss of lubricant/air boundary," is common for systems that have any form of seals, bearings, and lubricant. The lubricant can escape from the local bearing or the environment can become breached by dirt or other external contaminates. Thus, the causes of losing lubricant/air boundary describe the various ways lubricant can leak out or other things get inside.

Maintaining wheel roundness, 50500, is a subtle, yet important function of the subsystem. Without round wheels, a bicycle isn't worth much. The failure modes given for this functional failure represent some of the ways a wheel can be out of round. You can probably think of others.

Subsystem 60000: Rear Wheel and Tire

> **Description:** This subsystem accepts all acceleration and deceleration forces for transfer to the external system surface. The system effect is speed changes or speed maintenance. The spokes, suspension, and bearing characteristics are the same as the **Front Wheel and Tire** subsystem.

The explanations for this subsystem are omitted to help you think through the RCM functional details on your own. After going through five of the six bicycle subsystems, you should be ready to see the forest, not just the trees, or in other words, the functions rather than the equipment. Remember, the bicycle's subsystems fit together in the elegant compatibility that every good system enjoys. Use the results from the other subsystems to assist you in your analyses.

Identify Subsystem Interfaces

In Interfaces
Acceleration forces from **Power Transfer**
Deceleration forces from **Braking**
Operator weight forces from **Support and Stabilization**

Out Interfaces
Vertical and horizontal tire forces
Response forces to **Braking**
Response forces to **Power Transfer**
Response forces to **Support and Stabilization**
Operator weight forces from **Support and Stabilization**

Define Subsystem Functions & Functional Failures.

Functions
Accept acceleration forces from **Power Transfer**
Accept deceleration forces from **Braking**
Transfer rotation into linear motion
Provide external force damping to wheel
Maintain lubricant/air boundary
Maintain wheel roundness

Exhibit 6-12
RCM Spreadsheet for Rear Wheel and Tire at
Functional Failure Level

Functional Failures

60100	Unable to accept acceleration forces
60200	Unable to accept deceleration forces
60300	Total loss of rotation capability
60400	Partial loss of external force damping to wheel
60500	Total loss of external force damping to wheel
60600	Loss of lubricant/air boundary
60700	Loss of wheel roundness

Define Failure Modes for Each Functional Failure

Exhibit 6-13
RCM Spreadsheet Component for Rear Wheel and Tire

Index	Description
60100	**Unable to accept acceleration forces**
60101	Gear interface from **Power Transfer** failed
60102	Gear spokes sheared off
60200	**Unable to accept deceleration forces**
60201	Brake interface failed
60202	Gear spokes sheared off
60300	**Partial loss of rotation capability**
60301	Bad hub bearings
60302	Foreign object between wheel and **Operator Stab.**
60400	**Total loss of rotation capability**
60401	Major hub bearing failure
60402	Foreign object between wheel and **Operator Stab.**
60403	Hub bolt failure
60404	Major spoke failure
60500	**Loss of external force damping to wheel**
60501	Slow tire leak
60502	Fast tire leak
60503	Rim dented
60600	**Loss of lubricant/air boundary**
60601	Hub seal failure
60602	Hub bearing failure
60603	Hub covering puncture
60700	**Loss of wheel roundness**
60701	Dented rim or wheel
60702	Major spoke failure

An analogy to doing this part of the RCM method (Steps 1, 2, and 3) is building the frame, siding, and roof of a house. A logical, function-based model of the system has been built. The remaining steps are analogous to completing the interior. In RCM, this means considering the importance of the failure modes, matching modes with maintenance tasks, and delineating the labor resources required to put the plan into action.

There is no way to fully comprehend the power and functionality of the RCM method by just talking about its attributes. You must either perform a RCM study yourself or be guided, step by step, through one. While the method is versatile, it requires commitment and a skill that is hard to fully realize by anything other than experience with the level and type of detail required. If you have read through each of the six subsystems, I think you now agree that:

- The method can be applied at different levels of detail.
- The potential exists to become focused on details rather than at the level where maintenance can intervene.
- Once an RCM analysis has been performed, it has the potential for application to other systems with similar functions.
- The procedure could become tedious.
- You will never view the one speed coaster bicycle with the same blissful innocence again.

Categorize Maintenance Tasks

Up to this point, the RCM method has dealt with how a system is functionally structured. I am sure that you observed, and probably felt a little frustration in these discussions over the fact that some failures are clearly more serious than others. It is now time to address this concern. This is what maintenance task categorization is all about. An importance or criticality will now be assigned to each failure mode. The "importance" property is divided into four criticality classes: Safety, Operations, Economic, and Failure Finding. These four divisions help identify the necessity and justification for maintenance actions to mitigate the expression of failure modes. The more critical the failure mode, the higher the necessity and justification of assigning appropriate maintenance actions. To some extent, the criticality class also helps determine the type of maintenance action that is required.

The procedure for maintenance classification uses decision tree logic. This almost automatically classifies tasks by having the RCM analyst answer a series of simple binary (yes or no) questions. In general, you will want to make up your own decision logic tree, but for the purpose of illustration, we will use the tree shown in Figure 6-4 (Figure 5-5 repeated here for convenience). The classification process uses the decision tree down to the dotted line shown in this figure. The blocks below the line are addressed in the next section, which describes Step 5, the final step in the RCM method.

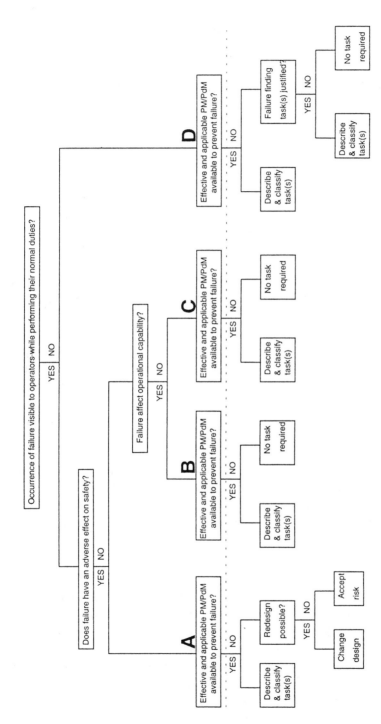

Figure 6-4. RCM decision tree example. (Copyright © 1986. Electric Power Research Institute. EPRI NP-4795. *Use of Reliability-Centered Maintenance for the McGuire Nuclear Station Feedwater System.* Reprinted with permission.)

Assigning Criticality Classes

The boxes labeled A, B, C, and D in the decision tree represent each of the criticality classes we will discuss in this example. Before we start to classify failure modes, let's discuss the meaning behind each criticality class.

Safety: Criticality Class A

Failure modes in this class are not usually difficult to identify. The question in the decision tree logic states its meaning: *"Does the failure have an adverse effect on safety?"*

This is not to say that everyone will agree on which failure modes should be assigned to class A. Here's an example. In a certain refinery during an RCM analysis, people from operations and maintenance met to review failure mode criticality assignments. When failure modes involving leaks of toxic and flammable gas products and by-products were addressed, operations always gave them a safety class assignment. From an operator's perspective, these failure modes were very, very dangerous. Operators worked in the local equipment environment and experienced the failures *without warning.* They had to react quickly and decisively to protect themselves with safety gear or make a fast exit. Maintenance personnel, at least initially, disagreed with the safety classification because they did not view these leaks as serious. From a maintenance perspective, they were correct. When maintenance personnel were called to fix the problem, they were responding to a *known hazard.* They took the appropriate safety measures *before* walking into the exposed area. With their safety training, they could work in the toxic environment safely. Consequently, these failure modes were not viewed as serious from a maintenance perspective. This example illustrates why a team approach to RCM applications is essential, because it helps to identify the different "standards of value" involved with the operation, maintenance, and management of the system under study.

Operations: Criticality Class B

If the failure under consideration does not have an adverse effect on safety, then the following question should be asked: *"Does the failure affect operational capability?"* A Yes answer to this question puts a failure mode in this class. The classification is usually applied to failure modes that directly affect operations. Make sure that operators who are involved with the study have a clear understanding of this question's meaning. Everyone involved must be clear on the precise interpretation of each question.

Economic: Criticality Class C

If the failure is not related to safety or operational capability, then the decision to assign preventive maintenance becomes a purely economic one. For example, let's consider the pumping configuration shown in Figure 6-5 (repeated here from Figure 5-1).

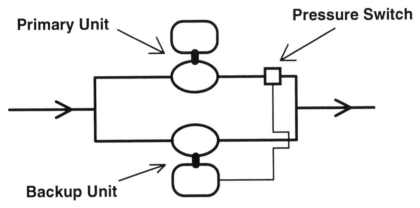

Figure 6-5. Inherent system redundancy.

If the primary motor-pump unit fails, the standby unit will automatically kick into operation. There will be no effect on either safety or production capability from the failure as the functional redundancies in the system are designed to handle it. Now a decision must be made about when to repair or replace the primary motor-pump unit. Factors such as the availability and cost of labor and parts must be weighed against the potential for failure of the standby unit, because such a failure would cause production to stop. Even though the class C category is two levels deep in the logic tree, failures in this class can still be of major concern and warrant serious attention. The point is that the importance of failure modes does not necessarily diminish as you move down the logic tree. The only characteristic that does change is the type of hazard produced by the failure mode.

Hidden: Criticality Class D

Failure modes in this class require special consideration. Notice in the decision tree of Figure 6-4 the first question is about this class: "*Is the occurrence of failure visible to operator while performing normal duties?"* The question is first because unless the operator knows a failure has occurred, there is really no reason to ask if the failure affects safety, operation, or economics. Such questions are only applied to failures that can be observed. In practice, you may want to ask these questions of hidden failures, too, in order to refine the classification categories. Most of the time, hidden failures are treated separately because of the unique characteristic that they are unknown to the operator at the time of failure occurrence.

Hidden failures were discussed in more detail in Chapter 5. That discussion is summarized here by using an example. Consider the pumping configuration shown in Figure 6-5. If the primary motor-pump unit is running, a failure in the standby unit would be a hidden failure. The operator would not know the standby unit has failed until it is needed.

Usually hidden failures occur in standby and safety systems that are supposed to operate only when certain conditions are achieved. As you can see, these types of systems are very important *when* they are needed. At all other times, they are essentially non-existent in the process. How maintenance is performed for these special systems requires special planning. This is why hidden failure modes receive special attention.

To demonstrate the use of the decision tree, let's go through a subsystem and apply it to each failure mode. Remember, the tree is to be applied to failure *modes*, not functional failures. Failure modes are events or sequences of events described at the level of detail where maintenance planning can be applied. Functional failures are not specific enough to be able to apply maintenance procedures.

We'll use the **Power Transfer** subsystem because it demonstrates all four failure mode classifications in a straightforward fashion. It will also be apparent that even when applying logic to a systematic, functional division of a system's operation, human judgment still plays a major role.

Below we consider the total subsystem failure mode listing for each identified functional failure, let's go through some failure modes with the decision tree logic. This exercise will help illustrate the thought processes that are used to make the classification decisions.

10100	**Loss of operator force transfer to pedals**
10103	Both pedals fail

Is the occurrence of failure visible to operator while performing

normal duties?: Yes

Does the failure have an adverse effect on safety?: Yes

Classification Result: A

10200 Loss of forward force to rear wheel
10201 Foreign object between chain and primary gear

Is the occurrence of failure visible to operator while performing normal duties?: Yes

Does the failure have an adverse effect on safety?: No

Does the failure affect operational capability?: Yes

Classification Result: B

10100 Loss of operator force transfer to pedals
10105 Poor pedal surface

Is the occurrence of failure visible to operator while performing normal duties?: Yes

Does the failure have an adverse effect on safety?: No

Does the failure affect operational capability?: No

Classification Result: C

10400 Lubricant/air boundary failure
10401 Primary gear axle seal failure

Is the occurrence of failure visible to operator while performing normal duties?: No

Classification Result: D

Exhibit 6-14 shows the criticality classes assigned to functional failures.

After looking at these classification examples you might wonder why answering such a seemingly simple set of questions warrants an entire step in the RCM method description. First of all, there are situations where answering these questions is not so simple. For example, in the first case regarding classifying failure mode 10101, if one pedal fails, you must study the operation of the bicycle under this failure mode and decide whether or not the failure mode adversely influences safety and operation. These type of decisions take time and sometimes considerable thought.

Second, the set of questions must be applied to each failure mode. There are more failure modes than any other element of the RCM method. It is common, in RCM studies, to spend more time analyzing failure modes than any other step. Even with the relatively simple, one speed coaster bicycle used in this example, there are 6 subsystems, 30 functional failures, and 94 failure modes.

Exhibit 6-14
RCM Spreadsheet Component for Power Transfer Subsystem

Index	Description	Criticality Class
10100	**Loss of operator force transfer to pedals**	
10101	1 pedal failure	B
10102	1 pedal bearing fails	B
10103	Both pedals fail	A
10104	Both pedal bearings fail	B
10105	Poor pedal surface	C
10200	**Loss of forward force to rear wheel**	
10201	Chain failure	B
10202	Primary gear failure	B
10203	Foreign object between chain and primary gear	B
10300	**Loss of backward force to rear wheel**	
10301	Chain failure	A
10302	Primary gear failure	A
10303	Foreign object between chain and primary gear	B
10400	**Lubricant/air boundary failure**	
10401	Primary gear axle seal failure	D
10402	Pedal seal failure	D

Implement Maintenance Tasks

If I had to choose a step of the RCM method that is the most important, it would be this one. Why? The first four steps constitute a plan, a paper study, or maintenance design. Up to this point, the actual maintenance being performed on the system has not changed. In this step, we'll develop the maintenance tasks that will be implemented to address the problems identified by each failure mode. In this next section, I'll describe one way to assign the tasks and frequencies and keep track of the labor resources that are needed.

Maintenance tasks for the one-speed coaster bicycle consist of four types denoted by the letters V, L, C, and R and defined as follows:

V Pre-ride visual inspection
L Lubrication
C Cleaning
R Adjustment

These are basically the types of maintenance that can be performed on a bicycle. They are directly related to the failure modes we have entered in the spreadsheet. For simplicity's sake, replacement and preventive maintenance tasks are deliberately not included here. In a real-life situation, they would be part of the process.

Let's talk about each type of maintenance:

The pre-ride visual inspection means the operator will visually check to see if these failure modes have occurred. This is part of the operator performed maintenance (OPM) philosophy. The visual inspection procedure encourages operator ownership of the system. Most people who ride one-speed coaster bicycles do own them. There is a naturally high degree of self-interest and self-preservation already present. Ideally, this feeling of ownership also applies in industrial settings.

Lubrication, cleaning, and adjustment tasks are self-explanatory. When applying this procedure to more practical and complex systems, the same general method of categorizing maintenance tasks for the equipment applies. The only difference is that there will be different types of task groups under the overall heading of the categories, V, L, C, and R. For example, different lubrication task groups could be labeled, L_1, L_2, etc. There are likely to be more categories than given here when predictive maintenance activities are included.

Four columns will now be added to the RCM spreadsheet to keep track of the frequencies of tasks assigned to each failure mode. The correlation of failure modes with available and pertinent maintenance tasks represents the first of two major milestones of maintenance task implementation. It depicts the RCM team's best judgment of how the system can be maintained. It is the connection between the systematic, logical, functional system division, the available resources, and the RCM team's creativity for how the resources can be best used together. The correlation of failure modes to tasks involves compromises. The RCM team's challenge is to distribute the resources most effectively to cover the most important failure modes. After this is done, if there are still important failure modes that cannot be assigned adequate maintenance tasks, the RCM project results can be used as evidence in the discussions about how best to accomplish the required maintenance. Options include more labor, more sophisticated predictive equipment, specialized maintenance task development, engineering re-design, or, in some cases, management may choose to accept the liability without any changes. Regardless of the outcome of such a discussion, the RCM methodology is a procedure that can be used to document changes in the number of resources required to efficiently maintain system function.

Let's return to the spreadsheet (Exhibit 6-14) for the coaster bicycle. With the criticality class assigned, maintenance tasks are now matched with failure modes. After each maintenance category, V, L, C, or R, the task frequency is given in terms of the number of times each year it should be performed. To illustrate this, suppose a failure mode has under the L column a frequency of three (3). This means that three times a year, or every four months, the parts are lubricated. Tasks that are either a part of the pre-ride inspection or part of the general operating procedures performed with each use of the bicycle are indicated with an asterisk (*). Before the complete subsystem is shown, let's discuss the portion of the spreadsheet shown in Exhibit 6-15.

Exhibit 6-15
Failure Mode Correlation With Available Maintenance Tasks

Index	Description	Criticality Class	V	L	C	R
10100	**Loss of operator force transfer to pedals**					
10101	1 pedal failure	B	—	3	2	—
10102	1 pedal bearing failure	B	—	2	2	—
10103	Both pedals fail	A	*	—	—	1
10104	Both pedal bearings fail	B	—	—	—	1
10105	Poor pedal surface	C	*	—	1	—

In this portion of the **Power Transfer** subsystem spreadsheet, certain decisions were made regarding which tasks fit which failure modes. Here is some of the reasoning that was applied. The inspection for the "Both pedals" failure mode automatically covers the single pedal inspections. The same reasoning was used for the lubrication and cleaning tasks. An additional adjustment task was applied to the "Both pedals" failure modes. This task is designed to tighten the pedal connections to the supporting structure and the supporting structure connections to the primary axle. The same task could be placed after each of the single pedal failure modes. In either case, the same amount of maintenance would be performed.

The last discussion may appear somewhat superfluous, because who really cares about the details of coaster bicycle maintenance? I included it, though, because it highlights a very important aspect of all good preventive maintenance programs. That is, one task can mitigate the expression of more than one failure mode. This should not be an uncommon occurrence. In general, maintenance tasks should be designed this way. Typically, though, maintenance programs have evolved in an unstructured manner, with changes in equipment, procedures, and available labor taken into account as equipment was added, failures occurred, or maybe in an annual re-evaluation of maintenance policy. The result? A lot of duplicate or overlapping maintenance and associated costs. The RCM method supplies a clear functional connection to each failure mode. It makes sense to use this structure as a guide for developing a maintenance design. Sometimes this happens automatically as in the case of lubricating pedal bearings; other times you have to be creative.

The **Power Transfer** maintenance task failure mode assignments are shown in Exhibit 6-16.

Notice in Exhibit 6-16 that functional failures 10200 and 10300 have exactly the same maintenance. If only 10200 existed, then considerably less maintenance would have been assigned. It is because of the "A" criticality class assignment for failures modes 10301 and 10302 that annual visual inspections were added. The severity of the failure modes is a major consideration for when and when not to apply maintenance. There may be failure modes in criticality class C for which no maintenance is applied. With limited resources available, it is possible that the RCM team would decide not to perform maintenance for certain failure modes. This action is most likely to be applied to Class C failure modes where the failure effects are economic rather than safety or production oriented. For example, certain motors may be selected to run to failure because the economic return on the amount of maintenance to prevent their failure is not cost-justifiable under the current operating conditions. If, on the other hand, there are safety-related failure modes that are not matched with suitable maintenance tasks, the liability of this condition may justify the additional costs of either revisiting the engineering design or applying specialized maintenance tasks. The RCM method provides this additional benefit, showing areas where more maintenance tasks are required, areas where

Exhibit 6-16
Power Tranfer Maintenance Task-Failure Mode Correlation

Index	Description	Critically Class	V	L	C	R
10100	**Loss of operator force transfer to pedals**					
10101	1 pedal failure	B	—	3	2	—
10102	1 pedal bearing failure	B	—	2	2	—
10103	Both pedals fail	A	*	—	—	1
10104	Both pedal bearings fail	B	—	—	—	1
10105	Poor pedal surface	C	*	—	1	—
10200	**Loss of forward force to rear wheel**					
10201	Chain failure	B	*	3	—	1
10202	Primary gear failure	B	1	—	—	⅓
10203	Foreign object between chain & primary gear	B	*	—	2	—
10300	**Loss of backward force to rear wheel**					
10301	Chain failure	A	*	3	—	1
10302	Primary gear failure	A	1	—	—	⅓
10303	Foreign object between chain & primary gear	B	*	—	2	—
10400	**Lubricant/air boundary failure**					
10401	Primary gear axle seal failure	D	—	1	1	—
10402	Pedal seal failure	D	—	1	1	—

maintenance tasks need to be changed, and areas where few or no maintenance tasks need to be performed.

Finally, we have to quantify the labor resources required. Usually, the amount of labor required under an RCM plan is less than current levels. (Discussion of some actual cases are given in Chapter 7.) One way of computing labor requirements comes from data given in the RCM spreadsheet. From this detailed listing of failure modes and maintenance tasks, the time and required labor can also be either input along with the task type and frequency information or approximated.

To show an example of how to estimate labor resource requirements from the spreadsheet, let's use the task failure mode listing in Exhibit 6-16. The spreadsheet excerpt is expanded to include a task number, craft required, and approximate task duration. This information is reviewed at the time when the maintenance tasks are matched with failure modes. If it is put into the spreadsheet at that time, then summarizing labor requirements can be accomplished without additional administration. The results can be obtained directly with the information in the spreadsheet and the functionality inherent in the software. An example of how the spreadsheet might appear is given in Exhibit 6-17.

The "Task #" field can incorporate the craft type and general type of maintenance task. In this example, only the task type is included. I think you can see how much information could be included in this field by adopting a code to categorize maintenance tasks. The operator-performed tasks are prefixed with an "O." The other task type applies the notation that was previously used. In actual maintenance studies, prefixes such as "E" for electrical and "M" for mechanical could be meaningful. The "Task Time" field shows the amount of time, in tenths of an hour, required to perform the indicated task. "Task Freq." shows the number of times a year the tasks are performed. This number fits naturally into the yearly labor resources calculation.

Our description of the RCM method appears finished. After all, what else could be described? We have gone through the five steps. What's missing? In a word - *Implementation.* Up to this point we have designed, built, and tested a maintenance plan. It still must be put into everyday practice. Implementation is everything! People must now change their behavior patterns and do things differently. They must understand that a system's functions and functional failures are at least logical. Trying to change even the smallest aspects of peoples' perspectives is the most difficult part of the RCM challenge. With implementation, labor resources must be re-allocated, and management practices must change. Without successful integration into the plant's everyday activities, the best RCM plan will be a failure. No book can tell you how to best implement RCM in your company. Every situation is different and requires the specific technical and political knowledge that is only known to those inside the corporate culture.

There are some factors that can help with successful implementation of RCM projects, though. In my opinion, the most important of these is across-the-board teamwork. This means having everyone involved with the project aligned and onboard, from management to the operations and maintenance staff workers. The actual RCM team must demonstrate a special type of cooperation because it combines the experiences of those on the equipment scene with the broad perspective and vision of the compa-

Exhibit 6-17
Failure Mode Correlation With Available Maintenance Tasks Incorporating Craft Type and Task Duration

Index 10100	Description Loss of operator force transfer to pedals	Criticality Class	Task #	Task Time	Task #	Task Time	Task Freq.	Task #	Task Time	Task Freq.	Task #	Task Time	Task Freq.
10101	1 pedal fails	B	—	—	L-13	.1	3	C-13	.1	2	—	—	—
10102	1 pedal bearing fails	B	—	—	L-14	.1	2	C-14	.2	2	—	—	—
10103	Both pedals fail	A	O-14	.1	—	—	—	—	—	—	A-14	.3	—
10104	Both pedal bearings fail	B	—	—	—	—	—	—	—	—	A-15	.3	1
10105	Poor pedal surface	C	O-15	—	—	—	—	C-15	.2	1	—	—	—

ny's direction-setters. In addition, hefty doses of time, patience, and perseverance are required as RCM's thoroughness sometimes makes it a laborious and slow process. Last but not least? Communication! There must be open communication between the RCM team and all parts of the corporation potentially affected by its implementation. Understanding the motivation, benefits, and effects is key to accepting change.

CHAPTER 7

Reliability-Centered Maintenance Case Studies

There is no security on this earth, there is only opportunity.

General Douglas MacArthur

Let's face it, change is a dramatic process. To override the momentum of the *status quo* and the security of the familiar requires a lot of work, confidence, vision, and a little bit of faith. It is always helpful to learn of other industries' experiences with the whys, hows, and effects of change. This is what case studies are all about.

Facing the effects of change is somewhat like standing at the water's edge on the first hot day of the summer. Regardless of how many initially get wet, there is almost always at least one cautious person who asks "How's the water?" Knowing how cold or hot the water is helps you to prepare for the initial shock or pleasure of getting wet. This chapter represents the initial swimmer and gives several different answers to the "How's the water?" question. It is a compilation of actual RCM study findings highlighting the costs, the savings, and the lessons learned. The following discussions and case studies are intended to help you "test the water" in your specific water hole. I hope the different segments give you a more secure sense of the viability of changing maintenance management to an RCM framework.

Where Should RCM Be Applied?

Selecting the system that is most suitable for the RCM analysis is not a trivial job. The challenge is to select a system that is large enough to make the analysis meaningful, yet small enough to make the job less than monumental. Independent of the size question is the concern for which system

158

will return the most benefits for the project cost incurred. If this question can be answered, system size is probably not a factor.

Here are some general guidelines for system selection. The systems chosen for analysis should satisfy one or more of the following criteria:

1. System contributes significantly to plant availability
2. Economic gains are expected from RCM
3. System failures affect safety
4. System is new and requires maintenance plan
5. There are regulatory/legal concerns
6. System failures carry environmental risk
7. System has had high proportion of labor and/or down-time
8. Data are available

There is no uniform method or formula for system selection that covers all situations. The priority given to the selection factors depends on each company's special situation. The factors listed are the most common issues from which the selection decision is made. Of this list, only data availability should not generally be considered a primary factor as a high degree of detailed historical information is not mandatory to support a useful analysis.

Because RCM can be applied in different forms, the method can be applied to systems other than just the high-profile, high-return systems. RCM can be applied to the entire plant in one way or another. A plant's systems could be allocated different degrees of detail in applying RCM to an entire plant. For instance, the most important systems can receive a detailed version, systems of moderate importance a streamlined version, and for the balance of the plant, RCM can be applied by only considering the philosophy in developing or reevaluating maintenance procedures and frequencies. For instance, it has been estimated for utilities that it would be cost beneficial to perform detailed RCM studies on just 20% of the systems. A streamlined RCM approach is applicable to 50% of the systems. The remaining 30% would receive an experience-based review applying the RCM philosophy [1].

Common Results of RCM Studies

At the maintenance task level, five types of actions can be taken:

1. Change time-based tasks to condition-based
2. Change current task content and/or task frequency
3. Add new tasks

4. Delete tasks

5. Develop engineering redesign of certain equipment or components

We'll review each of these actions before we go on to the specifics of the case studies.

Change Time-Based Tasks to Condition-Based Tasks

This is the most common result of RCM studies. It is partly due to the RCM method and partly to advances in predictive technologies. RCM studies survey many detailed failure modes and force a reevaluation of when, how, and why maintenance is performed. In many cases, these questions have not been asked regarding maintenance tasks in a long time, if at all. A contemporary review of failure mode-task correlation often reveals that the best way now to prevent certain failures is condition-based maintenance with the application of predictive measurement technologies.

Change Current Task Content and/or Task Frequency

Systems, subsystems, equipment, and components are constantly undergoing changes. The common term for these changes is "aging" and maintenance actions must change to reflect the "age" of the equipment. This does not necessarily imply that more maintenance should be performed as time advances. It means that system maintenance should change in time to reflect equipment and system condition and perhaps the changing importance of the equipment in the overall process flow of the plant. For example, the addition of emergency standby or other backup systems can change process flow pathways and influence system performance. Virtually any change in the equipment warrants a review of maintenance procedures. Another often overlooked aspect is that changing task frequency can affect the number and type of spare parts required to be kept in inventory. If task frequencies using these parts are extended, one direct, immediate effect achieved is the cost savings due to the reduction of inventory requirements.

Add New Tasks and/or Delete Existing Ones

Over an extended period of time, plant systems are altered to improve throughput, attain higher levels of up time, or take advantage of improvements in technology. If the changes are significant, then maintenance tasks are usually altered to reflect the new environment. However, even a series of small modifications can change the effectiveness of maintenance efforts. System improvements may also require the addition of new tasks due solely to the location or operation of new equipment. For example, suppose a pumping system is retrofitted with a standby motor-pump train, and the standby set is placed next to the primary train. Maintenance tasks must be developed to prevent failure modes where failures in one train cause failures in the standby train. These failures could be from major pipe breaks, explosions, or other relatively violent failures where the energy released from the failure would affect the operation of the second train.

Redesign Certain Equipment or Components

If failure modes are identified for which no effective maintenance tasks can be identified, there are two possible choices: accept the hazard associated with the failure (if any) or redesign the subsystem to eliminate the failure mode all together. Failure elimination is clearly the ideal, but it is difficult to achieve in practice. Design changes will have as a goal, at the very least, to move the criticality class of the failure mode out of the safety category. Other design changes must be justified on an economic basis. For operations-related failures, you usually can calculate or estimate the payback time. Non-operations related failure modes are those that do not directly affect production but have secondary consequences. For example, those failures that cause environmental releases whose costs are in the form of government imposed fines, cleanup, and public image are non-operative. Redesign decisions here are often difficult because the return on investment is not always obvious. A company can choose to deal with these scenarios with insurance rather than to directly try to prevent the failures. This decision is frequently the only real choice because failures are always possible for situations not contained in the analysis. It is never possible to design out all failure potential of any system. How much is enough?

Case Studies

The following sections present the results of various RCM projects in U.S. commercial nuclear power and other industries. As mentioned before, there are many variations of RCM. The projects are described here to give you a first-hand look at how others have applied the RCM method and what their experiences were like.

Case 1: Palo Verde Nuclear Generating Station [2]

Objectives:

1. To ensure the highest reliability possible for systems and their components
2. To identify maintenance for low-frequency failures
3. To reduce costs.

Description: In Palo Verde, Arizona, beginning in 1988, reliability-centered maintenance evaluations were performed on nine systems. They were the (1) emergency diesel generator, (2) the auxiliary feedwater system, (3) chemical and volume control, (4) essential spray ponds, (5) safety injection, (6) fire protection, (7) main feedwater, (8) essential cooling water, and (9) nuclear cooling water.

Costs:

2–3 man-months per system to develop
1–2 man-weeks per year per system to maintain

Benefits:

1. Maintenance time was reduced. For example, 1704 man-hours per year were saved through RCM implementation on the first pilot system, nuclear cooling water. 6,000 man-hours per year were saved on the second pilot system.
2. Spare parts inventory was reduced largely due to replacement of time-directed tasks by condition-monitoring tasks. This also reduced overhead associated with the parts (storage, ordering, paperwork, etc.)

Lessons Learned:

1. Prioritization/Dedication: The project must be given priority status within all organization involved. Resources from each area must be committed for the duration of the project. This includes full-time work from those directly involved in evaluations, as well as sufficient resource time from support areas.
2. Process: Use the version of RCM that best suits your needs. In this case, the team developed a stream-lined version called "Evaluation Logic." If outside information is available, use it to assist in developing your personal maintenance program. For example, this group

used standard preventive maintenance programs as a framework for developing their personal programs.

3. Pilot Projects: Choose the pilot program carefully, because it will be the basis for perceptions about the project. Keep the pilot small. In this case, the first system required the addition of a significant new task, resulting in only a small cost savings.

4. Scheduling: Develop your RCM implementation schedule after the pilot projects are completed in order to be most realistic. Consider the available resources, quality of tools, and depth of research you intend when scheduling the projects.

5. Documentation: Use common terminology in documentation. Be sure documentation is extensive enough to satisfy required regulatory scrutiny.

6. Data: Examine the quality and quantity of data at the beginning of the project. In this case, the group found a large quantity of questionable quality data. For the project, they worked almost entirely from paper reports.

7. Implementation: Assign tasks functionally, not according to organization. Greatest savings will be realized when personnel are trained on an extended range of tasks so that one person can perform several tasks while he/she is working on the equipment.

Case 2: Florida Power & Light: Component Cooling Water System [3]

Objectives:

1. To optimize preventive maintenance tasks
2. To reduce corrective maintenance
3. To save money

Description: The nuclear component cooling water system at FP&L's Turkey Point facility was the subject of this RCM project. Using the basic RCM methodology, 66 separate functions were identified, involving 180 functional failures. In total, 350 dominant failure modes were determined, 50 of which were considered major.

Costs: RCM for this complex system was completed over a two-year period.

Benefits:

1. Overall cost reductions for PM labor and materials expected to be 30 to 40%. Some of the reductions were due to several labor intensive time-based tasks converted to condition-based. In other instances, task frequencies were reduced.
2. Although not immediately measurable, it is expected that the new maintenance plan will reduce the number of outages, which cost approximately $700,000 per day.

Lessons Learned:

1. Success! The RCM methodology works.
2. System selection: RCM is not necessary for every system. Candidates for analysis should be those with extensive preventive and/or corrective maintenance histories to obtain maximum benefit for costs.

Case 3: R. E. Ginna Nuclear Station Maintenance Improvement [4]

Objectives:

1. To increase maintenance effectiveness for an aging plant
2. To ensure continued safe operation in a cost-effective manner
3. To prepare for licensing renewal
4. To optimize the ratio of preventive to corrective maintenance
5. To extend the life of the plant

Description: As Rochester Gas and Electric Corporation's Ginna nuclear power station approached its 20th anniversary, RCM was made a focal point in the facility's overall maintenance improvement program. Twenty-one systems were analyzed in the project, using standard RCM techniques. Labor was divided between contract consultants, who collected and analyzed data, and RG&E staff, who were responsible for technical review, recommendations, and implementation of the RCM plan.

Approximately 1,300 changes were made in maintenance task requirements.

Costs: Just over 1,250 labor-hours per system, at an average cost of $80,000 per system, for 21 systems. This cost included contract labor, and participation in industry group activities for technology transfer.

Benefits:

1. Objectives were met: A successful maintenance program was begun, addressing the need for an optimal PM plan to counter the effects of aging on the plant. This program will be continually modified, and is considered a "living" RCM program.
2. Cost savings: Over $45,000/year will be saved in labor costs alone as a result of changes to time- and condition-directed tasks.

Lessons Learned:

1. Background: Plant history must be well reviewed to discern local factors affecting equipment and its failures.
2. Support: A broad base of support for the project is needed to successfully develop and implement maintenance changes.
3. Commitment of resources: Team should be continuous throughout the duration of the project.
4. Prioritization: Implementation of RCM tasks must be given high priority.
5. Time requirements: Steps in the RCM analyses took approximately the following percentage of overall time:

System selection	2%
Data collection	3%
History analysis	24%
Failure mode effects/ Criticality analysis	30%
Fault tree modeling	40%
Plant life extension review	1%

Case 4: PSE&G Artificial Island Units: Streamlining RCM [5]

Objectives:

1. To use maintenance resources most effectively
2. To improve reliability

Description: RCM was the center of an extensive PM overhaul project at Public Service Electric & Gas of New Jersey. The projects began in late 1986 with classical RCM methods. Although two years of work produced high quality results on paper, implementation was virtually impossible. At this point, changes were made to use a more streamlined form of RCM, and to ensure implementation.

Costs/Benefit:

1. Approximately $5 million per year is saved as a result of the new PM program. This is as a result of reductions in 50–60% of the tasks that were analyzed.
2. One man-year is saved, on a year-to-year basis, for every two years spent in the initial analysis.
3. Other factors, such as increased safety, availability, cost of product, and reduction of failures will be measured over time.

Lessons Learned:

1. Process: As mentioned, classical RCM required a great deal of effort and resulted in high quality work. After two years, the group stream-lined the process in the following ways:
   - Reduced the amount of documentation
   - Focused on important failures of critical systems
   - Stopped building reliability models
   - Used relative values for failure effects, rather than completely quantitative values. This was particularly needed when little data was available
   - Excluded instrumentation
2. Tools: PSE&G developed a powerful set of computerized tools that turned out to be inflexible, particularly for logic trees and decision analysis. After the first two years, its use was limited to data gathering, documentation, and interface with their computerized maintenance management system, and performed most analysis on paper.
3. Implementation: Initially, maintenance personnel could not implement RCM recommendations into their busy day-to-day activities, particularly as many of the changes were described at a fairly high level. To put the new process into practice, the following changes were made:
   - Prioritization: The RCM team was supplied dedicated workers and given authority to implement recommendations.
   - Integration: Recommendations were restated as specific tasks, fully integrated into the existing maintenance programs.

- Measurement: Results of the changes are measured using "Impact Analysis" and are expressed in terms of hours and dollars saved.
- Communication: All personnel involved with the changes are kept informed.

Case 5: Quantitative RCM at Wisconsin Electric Power Company [6–7]

Objectives:

1. To reduce maintenance expenses without sacrificing safety or availability
2. To determine need for new predictive maintenance technologies
3. To optimize PM tasks in order to pay for new PM technologies

Description: The Wisconsin Power & Electric Company selected the Pulverized Coal System on its Oak Creek Units 7 and 8 for RCM analysis because it was identified by the CMMS as the system with the highest overall number of maintenance hours. As six years of failure data was available through the CMMS, a quantitative approach was taken, incorporating statistical analysis, to develop a highly cost-effective, ongoing maintenance program.

Costs: $32,000 or 800 man-hours for analysis

Benefits: Anticipated $177,000 per year

Lessons Learned:

1. Team size: An RCM project can be successfully completed by a team as small as one person with good analysis and communication skills.
2. Objectives met: The RCM study reduced the costs of maintenance without degrading performance. It also justified the addition of four new predictive maintenance technologies.

Case 6: RCM for Chemical Manufacturing [8–9]

Objectives:

1. To prevent vinyl chloride monomer (VCM) leakage and fires
2. To prevent pump system breakdown

3. To reduce the production deficit by improving system reliability
4. To reduce overall maintenance costs by focusing resource application

Description: VCM is a highly flammable, liquid carcinogen. VCM leaks of any size cannot be tolerated. The RCM study centered on a highly complex system chemical reactor charge pump system. The study was started because an accident had occurred. One of three motor-pump units started to leak VCM, resulting in a fire. A review of the pumping system maintenance history indicated that the pumps required rotor sleeve bearing changes every 30 months. Other than this one task, no other predictive or preventive maintenance tasks were documented. The failure that resulted in a fire consisted of multiple components. Two of the major factors in the accident sequence were failures of the motor winding and the failure of its circuit breaker. The accident consequences, including production lost, repair costs, cleanup costs, coupled with the apparent lack of suitable accident preventive maintenance, and the composite of the nature of the system failure, provided strong motivation for a critical assessment of the entire VCM pumping system and the suitability of its sealed or "canned" motor-pump units.

Two failure trees were used, one for hidden and one for visible failure modes. The hidden failure mode decision logic tree was used for failures of standby automatic protective devices, such as the circuit breaker that was found to have failed in this situation. Considerable effort was given to the analysis of such devices because they are often the last element of an accident sequence. If they operate, no accident occurs; if they fail, an accident and probably fire will occur.

Costs: Approximately six person-weeks (six people, one week) for the bulk of the analysis.

Benefits: A PM plan was created that addresses the problems and needs of the system.

Lessons Learned:

1. Team size: A team comprising representatives from all/most areas involved in the RCM project is needed to quickly complete analysis of large, complex systems. In this case, the RCM team comprised the machinery engineer, process design engineer, electrical engineer, a part-time maintenance supervisor, the instrument and control engineer, and a part-time operations supervisor.

2. Time: Given the right resources, RCM analysis can be successfully completed in a small time frame (in this case, one week).

3. Process: Analysis of the safety problems that lead to the RCM study clearly showed the interdependence of many electrical, mechanical, environmental, and procedural components of the complex system. RCM provided the procedure for systematically developing an understanding of these interactions. This study underlined the importance of designing maintenance at a functional level, considering the importance of each component at a system level.

Other RCM Experience

Our review of case studies showed both common and divergent practices, opinions, and results. RCM has been successfully performed for many reasons, with team sizes ranging from one to many, in time frames ranging from one week to over two years, and using many various approaches to the methodology. Your application of RCM must be tailored to your needs and resources. Here are some of the factors to be considered:

Urgency for RCM completion: An emergency situation like the one described with the VCM case obviously provides justification for pulling out all stops to get a plan constructed ASAP. Significant changes in procedures, regulation, or equipment, whether during a short time frame, or accrued over time since the initial maintenance plan was developed also increase the urgency.

Available Resources: This will include people and their time and commitment, and money.

What Can You Really Save? Although only quantifiable over a long term period, a number of projects specifically pointed out the large payoff that would be recognized by avoiding just a single failure.

Scope: How far into your systems do you really need to go, and with what level of detail? What can you do to recognize almost immediate savings? This is the place to start!

Data Availability: Although not a limiting factor for RCM, availability and quality of data will determine the approach you can take to RCM. Little data? Use a qualitative approach, with a common scale for comparison of failures. For situations where there is readily available data, a quantitative approach is more viable. The proliferation of CMMS will steer us in the latter direction in the future.

Support: Don't even bother starting an RCM project unless you have full management support and authority to implement the results. And let's not point the finger just at management. You have to communicate with everyone affected by the project, getting them involved as it affects them, or you will never get the support you need from those who can make change happen. You'll also learn a lot from them.

Implementation: Recommendations are cheap but only implementation counts. Up to the time of implementation, any RCM project is theory. Implementation represents a new and perhaps greater challenge than the RCM analysis. Even though the plant personnel agree with the RCM recommendations, who is going to take the time to change the recommendations into work orders, procedures, or design changes? The RCM project must be fully integrated into the current PM program and the maintenance work control process. Marketing on the project can't stop once the RCM project is started.

Concluding Remarks

I have tried to give you a concise look at how some companies have applied reliability-centered maintenance. As you might expect, each case study could be a book all by itself. Here are my concluding comments regarding how RCM studies will be performed.

1. Data availability is becoming less of a concern for many reasons. In some industries, regulations are requiring extensive record-keeping; in others accurate data is seen as the basis for increasing profitability. Regardless of the motive, the more accurate the data, the more precise the final RCM design. The proliferation of CMMS platforms has, and will be, a major catalyst for improving overall industrial data analysis capabilities. We can expect to see system and equipment functions integrated with system and equipment failures. Future generations of maintenance and operations information systems will better record and analyze RCM-based, functional information.

2. The challenge is to adapt the method so that an economic return is seen within one year. Originally, RCM was viewed as a multi-year, strategic direction to re-engineer maintenance design. Today's busi-

ness environment demands tactical solutions with almost immediate payback. Overall experience has shown the time frame for RCM completion is shrinking, as are the resources needed to complete a project. Companies are now looking at RCM as a method that has tactical returns AND provides a framework for strategic re-engineering. This is how RCM is evolving as it is adapted to satisfy the industrial marketplace.

The next chapters discuss the concept of risk and show how to integrate risk analysis with RCM. Why? Because reliability, profitability, and growth translate into business *security* as well as *opportunity.*

References

1. Toomey, G. E., "Full Plant Maintenance Program Optimization Using Reliability-Centered Maintenance," Nuclear Energy Conference, August 1992, pp. 267–272.
2. Davis, B., and Anderson, J., "Lessons Learned at Palo Verde Nuclear Generating Station," Proceedings of the American Power Conference, Vol. 53, part 2, 1991, pp. 1105–1110.
3. Smith, A. M., Vasudevan, R. V., Matteson, T. D., and Gaertner, J. P., "Enhancing Plant Preventive Maintenance Via RCM," ARMS, 1986, pp. 120–123.
4. Midgett, W. D. and Bettle, J. O., "Reliability-Centered Maintenance at Ginna: Results and Implementation," Proceedings of the American Power Conference, Vol. 53, part 2, 1991, pp. 1088–1092.
5. Strong, D. K. Jr., "Reliability-Centered Maintenance Streamlining Through Lessons Learned," Nuclear Energy Conference, August 1992, pp. 361–366.
6. Kowalski, M., A., Wisconsin Electric Power Company, private communication, January 1994.
7. Kowalski, M., A., "The Quantitative RCM Study of Oak Creek's Pulverized Coal System," ASME 93-JPGC-PWR-5, October 1993.
8. Pradham, S., "Applying Reliability-Centered Maintenance to Sealless Pumps," *Hydrocarbon Processing,* January 1993, p. 43.
9. Pradham, S., private communication, March 1993.

CHAPTER 8

On the Nature of Risk

Take calculated risks. That is quite different from being rash.

George Patton

Before we take even calculated risks, it makes sense to understand the concept of risk and the use of risk in decision-making. Here is a series of questions that will be addressed by the information in this chapter. You can answer these questions for yourself while you are reading the discussions that follow.

1. Is risk a "natural" metric?
2. Can risk be measured directly?
3. In what units is risk measured?
4. Can risk be added & subtracted?
5. Is *all* risk bad?
6. What exactly is meant by the term, "risk modification"?

Subjective or Qualitative Risk

The topic of risk has received a lot of press in the past few years. The terms "risk exposure," "high risk group," and "risk management" are all commonly mentioned in our society. It seems that the more we learn about our world, the more we learn how dangerous a place it really is! Our technological advances have enabled us to be aware of hazards that may or may not have always been there. We now know of more and more ways we can get injured, acquire a disease, and die. Choose your own examples. Most people feel they face more risk today than in the past, and they feel that risks will be higher in the future. This view, interestingly enough, is not shared by professional risk managers [1]. One thing is certain, it is not possible to avoid all risks.

Regardless of what definition you use, one thing is clear: *risk is bad*. The term is never applied in reference to good events. It is always associated with the likelihood of occurrence of undesirable and sometimes catastrophic occurrences. Whether it's the evening news, magazine reports, health reports, advertisements, wine bottles labels, cigarette packages, or gasoline cans, the term "risk" is always attached to things that are bad. It is common to hear or read about the *risk* of developing heart disease from smoking, but not about the *risk* of winning the lottery, as the latter is perceived as something good.

To most people, any discussion of the term "risk" is deemed philosophical, theoretical, and for the most part, irrelevant. It is a subject not related to their day-to-day existence. It is viewed as an esoteric tool of science, engineering, and political science—not as a "grass roots" metric. Risk is clearly used in these areas, but I suggest to you that risk analysis or risk measurement is just as common to your everyday existence as the concept of self-preservation. It is part of the human psyche.

Here's why. First of all, risk is always in the future. If something bad has happened, the risk of the event occurring no longer exists. Of course, depending upon the situation, the bad thing may happen again. Lightning can strike the same place twice, but risk always pertains to what *can* happen, not what *has* happened. This means that the more we know about what can happen, the more we can try to avoid the hazards. This may explain why most people perceive a greater risk than in the past. Our technological society is aware of more and more knowledge every day. This would also support why people perceive their risks as increasing. But we use information everyday to reduce our risks of our perceived hazards. We use seat belts, do not drive in severe weather, regulate our eating habits, avoid cigarette smoke, use smoke alarms, and so on for the purpose of reducing our *risk* of injury, disease, or death. This is risk analysis pure and simple. The human consciousness is rooted in self-preservation and inherently applies risk as a metric in planning its future course through the journey of life. Are there others? Sure; economics, time, emotions, and other such factors are all involved. The point is that risk analysis is natural and an innate characteristic of human existence.

Risk cannot be measured directly, but must be calculated. You cannot attach a probe to a device and measure its risk. From this perspective, risk is not a naturally occurring phenomenon. It is a parameter that requires the integration of at least two quantities: the chance and the type of event. This observation is extremely important. It means that risk is an abstract parameter requiring a degree of intellect that may perhaps be unique within the human species. Maybe the ability to perform risk calculations is related to who will succeed and who will not. A large part of intuition may be an unconscious, on-the-spot assessment of the circumstances at hand. Winning the lottery is luck, pure and simple, but success in any deliberate endeavor may be more dependent on risk assessment accuracy. The hypothesis makes sense to this humble risk analyst.

I wanted to discuss risk before I brought the "authorities" into the discussion. I feel their view is very clinical and too abbreviated to really add any clear insights in the beginning of this study of risk. It is time, however, to show how the experts in definitions define the word, risk.

Definitions of Risk

1. (1) Possibility of loss or injury: peril (2) a dangerous element or factor. (3a1) the chance of loss or the perils to the subject matter of an insurance contract. (3a2) the degree of probability of such loss. (3b) a person or thing that is a specified hazard to an insurer. (3c) an insurance hazard from a specified cause or source [2].
2. The chance of injury, damage, or loss [3].
3. (1) The possibility of suffering harm or loss; danger. (2) A factor, element, or course involving uncertain danger; hazard. (3a) The danger or probability of loss to an insurer. (3b) The amount that an insurance company stands to lose. (3c) A person or thing considered with respect to the possibility of loss to an insurer; *a poor risk* [4].

These definitions are clinical. If you are like me, you would say, "So what?" Looking at these meanings together does illustrate, in part, why people have different ideas of what risk is all about. The same general flavor of the meaning of risk is effectively portrayed by all of the definitions, but there is no one standard definition like there is in the definition of an object such as a Great Dane or baseball. In each definition above, the editorial committee has tried to give you an exact meaning of risk. The result is three versions of "exactness."

Subjectively, risk can be described as the perception of a hazard. How people view hazards will greatly influence their perception of an associated risk. It is a well-known fact that people generally are willing to accept higher risks if they feel they have some control over the process. For example, people feel fairly at ease driving automobiles, and yet feel threatened by living near a nuclear power plant, even though the actual incidence of death or injury caused by the former far, far outweighs the latter. However, in the case of the automobile, people have some control. With a reactor, they have none. In fact, one study suggests that people are willing to accept risks associated with voluntary activities 1000 times greater than risks associated with involuntary hazards! The study also suggested that people are willing to accept the risks associated with an activity roughly proportional to the third power of the activity's benefit [5]. How about that for risk quantification! The point I am trying to make does not lie in the value of these numbers; it is that risk assessment is a highly complex subject. It is almost guaranteed to invoke both sides of people's emotions. When dealing with subjective risk, my advice is: Be careful!

Pontius Pilate's question, "What is truth?" has a more modern form, "How safe is safe enough?" I am not even going to try to tackle that question. Other authors have addressed this issue, and I am going to pass the buck here. Developing an "essentially optimum balance" [6] between technological risks and benefits and offering foundations for developing consensus [7] have been very elegantly stated elsewhere.

This is in part why the topic of risk is such a political powder keg. Everyone is naturally different and everyone perceives hazards, and therefore risk, differently. It is hard to develop a rational consensus when people are not thinking along the same guidelines. A good illustration of this is shown in Table 8-1 [8]. Three different groups of people were asked to rank certain activities and technologies from the highest (rank 1) to lowest (rank 30) risk. Thirty college students, forty members of the League of Women Voters, and fifteen risk assessment experts made up three different groups. The results given in Table 8-1 are based on the geometric mean risk ratings within each group.

The point of this discussion is not the actual information content of Table 8-1. It is that the perception of hazards is highly subjective. It therefore is not surprising that risk assessment is a quagmire of politics, emotions, and individual value systems. The following quote indicates how much of a "muddy swamp" subjective risk really is [9].

"Research further indicates that disagreements about risk should not be expected to evaporate in the presence of evidence. Strong initial views are resistant to change because they influence the way that subsequent information is interpreted. New evidence appears reliable and informative if it is consistent with one's initial beliefs; contrary evidence tends to be dismissed as unreliable, erroneous, or unrepresentative [10]. When people lack strong prior opinions, the opposite situation exists—they are at the mercy of the problem formulation. Presenting the same information about risk in different ways (for example, mortality rates as opposed to survival rates) alters people's perspectives and actions." [11]

Risk assessment, in its wonderfully diverse way, is fundamental to how every person lives their life. Using an analogy on General Lew Wallace's

Table 8-1
Technology and Activity Risk Perceptions From Three Different Groups of People

Activity or Technology	League of Women Voters	College Students	Experts
Nuclear Power	1	1	20
Motor Vehicles	2	5	1
Hand guns	3	2	4
Smoking	4	3	2
Motorcycles	5	6	6
Alcoholic Beverages	6	7	3
General Aviation	7	15	12
Police Work	8	8	17
Pesticides	9	4	8
Surgery	10	11	5
Fire Fighting	11	10	18
Large Construction	12	14	13
Hunting	13	18	23
Spray Cans	14	13	26
Mountain Climbing	15	22	29
Bicycles	16	24	15
Commercial Aviation	17	16	16
Electric (Non-nuclear) Pwr	18	19	9
Swimming	19	30	10
Contraceptives	20	9	11
Skiing	21	25	30
X-rays	22	17	7
H.S./College Sports	23	26	27
Railroads	24	23	19
Food Preservatives	25	12	14
Food Coloring	26	20	21
Power Mowers	27	28	28
Prescription antibiotics	28	21	24
Home Applications	29	27	22

Adapted with permission from Societal Risk Assessment [8].

quote: "Beauty is altogether in the eye of the beholder," without a doubt, risk is in the mind of the perceiver.

Risk as a Quantitative Measure

We now turn the corner, leave the muddy swamp behind, and enter a world where logic and rational thought prevails. I am talking about a mix of applied mathematics and engineering. In this arena, risk has one, and only one definition. Here it is:

$$\text{Risk} = \text{Probability} \times \text{Consequence} \qquad (8\text{-}1)$$

The beautiful thing about this pseudo-formula is that everyone agrees on this definition of risk. In engineering and, to a growing extent, the industrial risk management community, this is risk—pure and simple. An alternative form is sometimes expressed as:

$$\text{Risk} = \text{Frequency} \times \text{Severity} \qquad (8\text{-}2)$$

This is not a different definition, but a version of the original. Both equations say the same thing. Probability is related to frequency and severity is the same as consequence. The version used depends on with which semantics people feel the most comfortable. Let's discuss each of these two quantities separately.

Probability

There are many ways to develop the probability of occurrence for certain events. Two of the major classifications used are discussed here. Keep in mind that probability is a branch of mathematics and is a field in itself. What follows here is a brief introduction to the subject.

Classic Probability

Probability describes the stochastic nature of the frequency of event occurrence. It is a mathematical quantity that can only have values between 0 and 1. It is defined as the number of desired outcomes divided by the total number of possible outcomes. There are many ways to calculate this quantity. Here are some examples, just to give a flavor of what a probability computation can involve. One fact that will be readily apparent is that you need data about the process to compute the probability of various outcomes.

Process: Tossing a Coin. The probability of getting "heads" in a coin toss is $\frac{1}{2}$; that is 1 desired outcome out of 2 possible outcomes.

Process: Rolling Two 6-sided Dice. The probability of rolling a seven using two dice is 1/6; or 6 desired outcomes out of 36 possible outcomes. Here are the desired outcomes:

Die 1	Die 2
1	6
2	5
3	4
4	3
5	2
6	1

There are 6 possible outcomes from the first die and six possible outcomes from the second die. The outcome of the second die is not influenced by the outcome of the first die. Therefore, the total number of outcomes is $6 \times 6 = 36$. The probability is computed as 6 desired outcomes divided by 36 possible outcomes or $\frac{1}{6}$. (Rolling a seven is the most likely outcome when rolling two 6-sided dice. This is why you loose when you roll a seven in the dice game of craps. Remember, the origin of probability has its roots in gambling.)

When you use probabilistic results you are gambling. Just because the probability of "heads" when tossing a coin is $\frac{1}{2}$ does not mean that heads will come up exactly 3 times in 6 throws. It only means that as the number of throws becomes larger, the number of heads will approach $\frac{1}{2}$ of the total number. Also, it is important to remember that regardless of how small a probability value, the event can happen very soon! This is exactly the case of the lottery winner who hits the jackpot with the first ticket they purchase. The converse is true for high-probability events. A high probability value does not guarantee the event will happen. In both cases, the probability results signify that on the average, given an infinite number of situations as the one you are in, the result indicated by the probability value will occur at the specified frequency. When you are dealing with situations where you only have one chance of success or failure, probabilistic methods may not be the best approach. But the practical reality in many industrial situations is that there is no other technique to assist in risk management. Probabilistic approaches are fine as long as people understand their strengths and limitations. And, as we saw in Chapter 3, there are many examples where the method is abused.

Bayesian Probability

The classic approach to probability that we just discussed is fine when sufficient failure or event data exist. However, the method falls short in situations where little or no data has been compiled. Consider, for example, a new plant where nothing has failed yet. Classic probability theory is not applicable to this situation. The Bayesian method can be used to compute failure probabilities, however, as it uses prior knowledge of failure rates from other sources, such as expert opinion, industry values, and other plant experience, along with plant-specific information as it becomes known. The result is called the posterior distribution and is computed from combining the prior and sample data, using a formula known as Bayes Theorem [12]. This distribution contains the probability values to be used in an application. The philosophy behind the Bayesian approach is that joining prior information with problem specific data yields a posterior distribution that is closer to the actual, ideal distribution than either of the two parts used in the calculation. Figure 8-1 shows graphically how a Bayesian method of computing probability works [13-14].

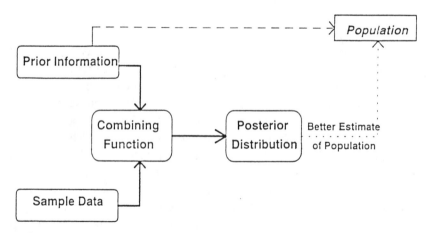

Figure 8-1. The Bayesian probability procedure.

Bayesian Examples

This example is discussed to give you some idea of the utility of this powerful technique. Its purpose is to illustrate one way Bayesian probability methods can work. The details of the mathematics are given in appropriate detail in other texts. Keep in mind there are many other application areas that range into virtually every facet of applied probability.

Suppose you are performing a quality study for an aluminum rolling mill where large ingots are trimmed and pressed into sheets of a thickness specified by the customer. A new rolling line just began operation, and an investigation is being performed as part of an overall cost-benefit study to judge the reliability of the new process line. Because the new line has just started operation, there is insufficient history to do a stand-alone evaluation of the line's quality performance. Previous experience indicates that the rolled aluminum sheet thickness can be modeled by a normal distribution function. In other words, the output product sheet thickness varies as a standard bell-shaped curve that is characteristic of the normal distribution. Analysis team members research other plants' experiences with this process and determine that the prior knowledge indicates the thickness has a mean, μ_0, and variance, σ_0^2.

With the new rolling line operating, quality data is collected as random samples of size, n. The sample mean, x_s, is computed as information that updates the prior experience results. The sample variance, σ_s^2, is assumed known. The Bayesian mean, μ_B, and variance, σ_B^2 of the resultant normal distribution incorporates both the prior knowledge $\{\mu_0, \sigma_0^2\}$ and actual data as it becomes known $\{ x_s, \sigma_s^2 \}$.

$$\mu_B = \frac{n * x_s * \sigma_0^2 + \mu_0 * \sigma_s^2}{n * \sigma_0^2 + \sigma_s^2} \tag{8-3}$$

$$\sigma_B^2 = \frac{\sigma_0^2 * \sigma_s^2}{n * \sigma_0^2 + \sigma_s^2} \tag{8-4}$$

Most of the time, it is not practical to assume that the sample variance, σ_s^2, is known. However, if random sample sizes equal to or greater than 30 values are taken, i.e., $n \geq 30$, the sample computed variance, s^2, can be used in its place.

Looking at the formulas above, you can see how the final distribution parameters $[\mu_B, \sigma_B^2]$ are composed from the prior data and process-specific data that is available. From a practical point of view, if you develop a "sufficient" database of plant failure data, you should be able to replace the Bayesian theory with classic probability. It is due to a lack of data that you are using the Bayesian method in the first place. The question of exactly when you should jump off the Bayesian methodology and use the classic approach has never been completely answered. Evidence does show that basically, with the proper treatment of data, both approaches should yield about the same results [15].

Here's another example that shows how previous data (prior distribution) and updated data (posterior distribution) are effectively combined. The mathematics of Bayesian analysis is not described here because, for this discussion, I am only demonstrating the method's power and versatility to handle practical problems.

Suppose you desire to compute the mean failure rate of an on-line system as a function of the number of hours of operation. Using the Gamma distribution to match this prior knowledge gives the mean failure rate (λ) and the variance (var λ) as shown on the left-hand side of Table 8-2 [16]. Now suppose we have additional, perhaps actual system data that indicates there have been "f" failures in "T" hours of operation. The right side of Table 8-2 shows the result of application of Bayes Theorem by combining this posterior, system-specific information in the mean and variance formulas. Notice that if the number of failures and total operation time, "f" & "T," are zero, the Bayes results collapse into the prior distribution results. Thus, you can use the Bayesian results as a continuous framework to measure failure rates (in this case) even if you have not (yet) experienced any failures.

Table 8-2 [16]
Demonstration of Bayesian Computed Mean Failure Rates

Prior(Before Updated Data)	Bayesian Results Incorporating Updated Data
$\lambda = \dfrac{\beta}{\alpha}$	$\lambda = \dfrac{\beta + f}{\alpha + T}$
$\text{var } \lambda = \dfrac{\beta}{\alpha^2}$	$\text{var } \lambda = \dfrac{\beta + f}{(\alpha + T)^2}$

Before applying this technique, you may want to first educate your audience and/or peers. Bayesian probability methods are not common in many circles and some people may be uncomfortable with using seemingly esoteric techniques. Bayesian methods *are* valuable. In many cases, they supply probabilistic information to enhance decision-making when no other method will work. Just beware that in the probability applications community, there are strong feelings both ways about the utility of Bayesian and Classic methodologies. If you use Bayesian methods, keep in mind you may receive some "non-believer" comments about the viability of the method. If you prepare ahead of time for these types of questions, their concerns can be put to rest without much trouble.

Consequence

Consequence denotes the magnitude of loss. It is somewhat subjective in the sense that the magnitude of loss can be viewed differently by different people and therefore can be a challenge to quantify. Within a plant or company, the best way to do this is by developing a consensus.

There is no standard way to calculate consequences. You cannot open a book and read the section on consequence estimation. In general, consequences describe how you can lose. In other words, these are the penalties associated with the occurrence of certain events. What are these penalties? They basically fall into the following categories. I will briefly discuss each one so that you can identify elements of each category that might be appropriate in your applications. As you will see, some of the descriptions overlap. This is because the consequences of events can affect more than one area at the same time.

Consequence Category #1: Safety

This category is by far the most important. Injuries and deaths caused by a system failure clearly have the most severe implications possible. The loss of a life or the pain of an injury are impossible to fully quantify. Yet even concrete costs such as worker's compensation and corporate liabilities are sometimes not fully taken into account in determining the implications of system failures. Why? Often the costs of these items come out of the corporate budget, not from the individual plant's budget. The consequence values are developed for plant use and not corporate or inter-plant applications. This practice is short-sighted and naive. Just as we must take a system-wide view of equipment to be truly effective, so must we realize that the health of any one component of a business will directly contribute to the success or failure of the company as a whole.

Safety-related system failures have other immediate implications. In the United States, corporate executives, plant managers, and even first-line supervisors are subject to civil and criminal penalties that can include large fines and even jail sentences if they are found to be negligent in a court of law. This sobering fact means that if a safety-related failure occurs and the judge or jury determine that you caused the failure, contributed to its causes, or were aware of the hazard and did nothing about it, you can be convicted and receive criminal penalties. This is not a forecast of the future. It has happened. The point I am making here is that in assigning consequence values to safety-failure events, make sure you take into account *all* of the factors that can affect the corporation, the plant, and you.

Consequence Category #2: Lost Production

If a machine that produces "widgets" fails, then lost production consequence can be either in terms of "lost widgets" or by factoring in costs in "lost dollar" terms. If a machine can make different grades of a product, such as a paper machine or a pipeline that can pump different products, then the computation of lost production in terms of dollars lost is much more complex. In this case, the cost of the loss will depend upon what product is being produced when the failure occurs. To further complicate the situation, just because a product with a higher market price is being made at the time of failure, this price may not be directly proportional to the actual costs of production. Different products have different profit margins that must also be considered in developing failure consequence values. There is no general formula that includes all of these factors. The point here is that you must be sure that you consider all of the important variables in your consequence decision-making when quantifying the costs of lost production.

Consequence Category #3: Lost Quality

Quality has evolved from buzzword status into a serious business-related concern. If a system failure affects the quality of a product, it produces a different, unique characteristic from all other failure consequences. It signifies that production does not necessarily stop, but that the nature of the product is not at the level either required or expected by the customer. How your customers react to this type of behavior from you, the supplier, determines how much emphasis should be placed on quality-related system failures factor.

Quality may or may not be a driving force in your business planning and operation. To some extent, most people now believe that quality is worth monitoring and improving. The field of business competition has grown to a world-wide arena. The delivery of product on time that meets or exceeds customers' expectations is becoming, and in some markets has already become, a requirement for business survival. Quality is no longer just a nice thing to have. It is mandatory. The bottom line: Consider product quality effects when developing system failure consequences.

Consequence Category #4: Environment Effects

This is a difficult category to quantify. The costs for cleanup of environmental spills is relatively easy to calculate. For the most part, so are the size of the fines that the company will incur as a result of damaging the environment. Because costs increase with the scope of the failure, the failure modes developed for each functional failure could be graduated to more accurately describe the potential implications of environmental damage from each of the potential system failures. Other consequences, such as the actual damage done to the environment, and even the effects of negative publicity surrounding environmental accidents, are far more difficult to put a price tag on.

Consequence Category #5: Maintenance Repair Costs

The maintenance repair costs category is the easiest to describe because the data required for description of the consequence are collected as a part of the usual business activity. You are likely to find, however, that some failure modes may never have been experienced and so no historical basis exists from which representative costs can be computed. Here are some options you can use to resolve this problem:

1. Solicit the experience of other companies. Depending how friendly you are with your competition, you might get some help there, but be aware that there may not be a direct correlation between the experiences of your companies as your cultures, organization, and accounting practices will vary. As examples, the other company may be unionized while your company is not (or vice versa), or they may incur different charges through the use of contract maintenance labor as opposed to company labor. In any case, even for information obtained from a professional society, seriously consider the source and composition of outside information. It is important to normalize these figures so that you are comparing like factors in like or similar conditions.

2. Solicit the experience of other plants within your company. Be aware that cultural, organizational, and accounting differences can occur even within a company and compensate for such differences in your calculations.

3. Get the people who know the equipment best together and estimate. Most simply this means you should apply corporate consensus-based expert opinion to form an estimate or "a reasonable educated mathematical guess" [17] of the maintenance repair cost.

Consistency in Consequence Values

To provide a consistent measure of risk, the units of all five consequence categories must be the same. You must be able to add risk contributions from each of the different sources, which you cannot do if they have different units. If your measure of consequence for each of the categories is not alike, you must convert all consequence values to a standard unit. The most natural units are dollars, because risk can be interpreted as the expected loss due to a certain event or a group of events. Therefore, losses are usually expressed in terms of dollars.

There is one conspicuous exception to using dollars as the unit for expression of consequence values, and that is in the modeling of the consequences of large toxic material releases. In this case, consequences are computed in terms either of concentration or, if public or human effects are under study, deaths and injuries. Clearly, the units of consequence are determined by the intent of the study. In the cases where human death, injury, and disease are used it is because people were directly and severely affected, and the human consequences are what really count in computing the risk of toxic environmental releases. The appropriate units therefore are human effects. To compute consequences in terms of dollars would require the computation of the worth of human life. Lawyers in civil court routinely perform this tragic and sad calculation on an individual-by-individual basis. In this type of consequence modeling what counts is how many human lives are affected and to what degree. That is why consequences are measured in terms of their human effects.

Risk Modification

The more we use the term "risk" in our day-to-day lives, the more people are changing their lifestyles to reduce risk exposures. Whether it is changing eating habits, quitting smoking, getting more exercise, or wearing a seat belt in a car, "risk modification" is an activity that everyone does to some extent. In this section, we'll discuss the nature of risk modification from a conceptual perspective. We'll first define the term and then talk about the various ways it can occur.

Risk is a function of time. It is also a function of many other things depending upon the subject, but time is the fundamental independent variable. Therefore in equation form, if $R(t)$ denotes the value of risk at time t, $P(t)$ the value of probability at time t, and $C(t)$ the value of consequence at time t,

$$R(t) = P(t) * C(t) \tag{8-5}$$

From Equation 8-5, there are three ways to modify risk:

1. Change the probability, $P(t)$
2. Change the consequence, $C(t)$
3. Change both the probability and the consequence

Risk modification is defined as the net change in risk over time from changes in probability and(or) changes in consequence. In mathematical terms, the time rate of change of risk, $\dfrac{dR(t)}{dt}$, can be written in terms of its dependent parts :

$$\frac{dR(t)}{dt} = \frac{dP(t)}{dt} * C(t) + P(t) * \frac{dC(t)}{dt} \tag{8-6}$$

Equation 8-6 shows that the fastest way to change risk is to change the *probability* of the largest consequence events and change the *consequences* of the highest probability events. This statement is not earth shattering and certainly does not require much mathematical insight. Equation 8-6 is a mathematical version of just plain common sense. This does not mean that it is without value. Equation 8-6 provides a quantitative structure for risk modification. Because risk is such a potentially volatile subject, this mathematical structure provides a mechanism to perform risk analysis on the common ground of analytical logic.

Risk modification is the change in risk over a specified time period, and indicated by the variable $\Delta R(t)$. Manipulating Equation 8-6 and allowing dt to be approximated by Δt, risk modification can be written in terms of changes in its constituent parts:

$$\Delta R(t) = \Delta P(t) * C(t) + P(t) * \Delta C(t) \tag{8-7}$$

Equation 8-7 describes how risk can be changed. Overall risk reductions correspond to $\Delta R < 0$ while risk increases occur when $\Delta R > 0$. Because probability and consequence are independent of one another, it is possible to have a zero net change in risk, i.e., $\Delta R = 0$, by having a risk increase in the first term, $\Delta P(t) * C(t)$ combine with a risk decrease in the second term, $P(t) * \Delta C(t)$. The most efficient and best way to reduce risk is by having both terms of Equation 8-7 negative. With this equation, we are able to see how risk changes in terms of its basic factors and establish a quantitative framework to compute risk modification.

The equations above illustrate how complex the subject of risk really is. The purpose of the mathematical discussion was to logically show the many variations in risk modification that are produced from the deceptively simple risk equation. Because there is more than one way to get the same risk modification value, in practice, which one is better? Are all ways of reducing risk effectively the same? Is there an optimum way to reduce risk? Because risk pertains to the future, can one way of reducing risk be judged better than others? The next section discusses answers to these questions by providing a geometric structure for representing risk measurement, modification, and assessment.

Geometric Representation of Risk

This section develops the risk coordinate system concept and discusses its characteristics to graphically show how risk can be represented, measured, and assessed.

Risk is a direct function of its two factors, probability and consequence. We can let them represent the coordinate axes in a common x-y plot configuration. In mathematical terms, the variables probability and consequence, and the subsequent risk value can be represented by a point in a two-dimensional Euclidean coordinate system. Every ordered pair of probability value and its corresponding consequence value, [p,c], represents one and only one point in this system. These facts allow the risk coordinate system to be useful by visually showing how risk can be represented, measured, and assessed. Figure 8-2 shows the risk coordinate system. In this example, the probability ranges from its theoretical minimum to its maximum possible value. This is fine. Consequence values, on the other hand, were arbitrarily set to a range of 0 to 100 for the purpose of this example only. In practice, consequence values will be customized to the problem. There are no standard values, maximum range, or minimum range.

Even though every point in the coordinate system has its unique representation [p,c], there are many points that can produce the same value of risk. For a straightforward example of this fact, suppose we plot the fol-

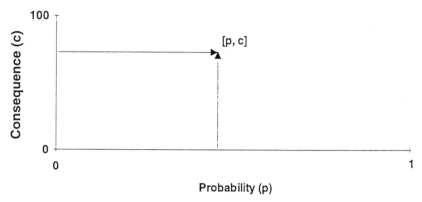

Figure 8-2. Risk coordinate system.

lowing two points that have the same product, i.e., same risk. Consider the following two points:

Point 1: [p = 1/5, c = 5]   Risk = 1
Point 2: [p = 1/2, c = 2]   Risk = 1

The result is shown in Figure 8-3.

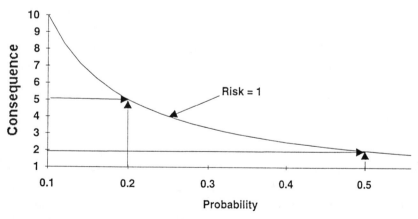

Figure 8-3. Properties of constant risk.

There are an infinite number of points that can produce the same value of risk. From the point of view of risk measurement, the situation where there are many failures (high probability) each with small consequences is equivalent to the opposite case where there are very few failures, each failure having much higher consequences. As Figure 8-4 shows, we can plot sets of points that all have the same risk value. This set of points is called an "iso-risk contour" or "risk contour" for short. They form a smooth curve in the coordinate system. Figure 8-4 also indicates that if probability and consequence are plotted in a standard linear format, the curves are nonlinear (i.e., not straight lines) in form.

When the same curves are represented by a log-log plot of probability and consequence, the risk contours become straight lines as shown in Figure 8-5.

If you are not familiar with the log-log plot format, it is just as easy to use as the standard linear version. Here's a brief review. Write the number you are going to plot in scientific notation. This means factoring out the powers of ten. Here are two examples:

General Format	*Scientific Notation Format*
0.000543	5.43×10^{-4}
5,430,000	$5.43 \times 10^{+6}$

A log scale has a repetitive set of lines. At the beginning of each set, write the power of ten that will be used for assigning numbers in each particular log set. Next, find each number's position on axis' scale inside the power of ten range you just labeled. Do this with the x and y (or in this case, the p and c) axes and plot the point. The spacing built into the plot automatically takes care of the logarithm for you.

Why would you use the log-log plot version? When the range of the numbers you need to plot spans more than one or two powers of ten, the linear scale becomes hard to read and usually gives more visible importance to the larger numbers. For example, suppose you had the following x coordinate points to plot:

$$0.001, 0.01, 1, 54.356, 1,023$$

The numbers range over six orders of magnitude. If you set the scale using the first two numbers on the left side of the sequence, you will need a piece of graph paper probably a mile long to include the last numbers. This clearly won't work. If you use the last two numbers to determine the scale, you will need a microscope to plot the first two numbers. In a log plot, each number gets equal billing with its own power of ten range for plotting.

This discussion of log-log plots is for a good reason. It is more than a matter of convenience or preference for dealing with straight compared to

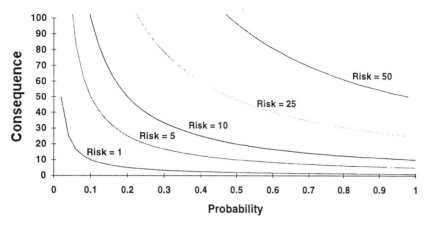

Figure 8-4. Contours of constant risk.

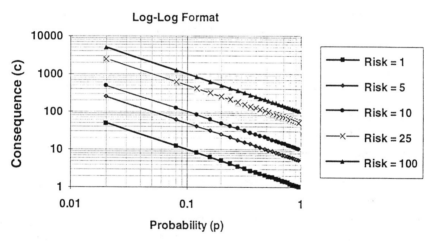

Figure 8-5. Constant risk contours.

curved contour lines. It is a necessity. Every realistic application of the risk coordinate system that I have seen has dealt with probability and consequence values that span several powers of ten. Table 8-3 gives an example of some fairly common intervals, showing the range of frequency values and their "occurrence" equivalents.

After reading Table 8-3, you will probably agree that the log scale makes sense to use. It is not uncommon to hear weather forecasters talk about "100-year" storms. In city planning, the "500-year-flood" is also used. For example, downtown Hartford, Connecticut is protected from the

Table 8-3
Comparison of "Occurrence" and Frequency

"Once Every x Years"	Frequency per Day
1 year	2.7×10^{-3}
10 years	2.7×10^{-4}
100 years	2.7×10^{-5}
1,000 years	2.7×10^{-6}
10,000 years	2.7×10^{-7}
100,000 years	2.7×10^{-8}
1,000,000 years	2.7×10^{-9}

Connecticut River by a "500-year" concrete flood wall. Statistically speaking, on the average every 500 years, the wall will fail to protect the city from a devastating flood. So far it has worked. In risk analysis and when dealing with low-frequency events, it is also common to deal with events that are postulated to occur every 10,000 years, 100,000 years, or more. Remember, just because an event has an extremely low probability or frequency, it can still happen today, tomorrow, or next week. In fact, it could occur today *and* tomorrow *and* next week. It is not likely, but neither are any other catastrophes that strike without any apparent schedule.

Now that you are satisfied, I hope, that probability can range over many orders of magnitudes, you might ask the question, "How infrequently must an event happen to be considered "irrelevant?" In practice, drawing the line to the lower limit of probabilities considered by a risk analysis is a part of the study. Remember, risk is mostly a relative quantity. A risk value only makes sense when compared to values from other events or the same event over a different time interval. Absolute risk estimates or risk standards are used today in measuring acceptable risk in some government and international cases. For the most part, though, risk values estimates are used for relative comparisons. In any given study, the lower limits of probability are often a natural outcome of the work. This will be seen in Chapter 9. In other cases, the risk analyst or analysis team can set the lower limit in line with an external constraint, such as available resources, or by comparing the risk to natural disasters like large meteor strikes in the facility.

Let's return to the risk coordinate system and continue to study its utility. The discussion so far has developed how risk can be visually plotted.

What purpose does the visual representation serve? Risk can be represented in the risk coordinate system, and risk reduction can be seen if the recomputed risk is on a smaller risk contour line. The main purpose of the risk coordinate system is to assist in the development of risk assessment conclusions for the process, plant, or event under study. Risk assessment is the process of judging the accident potential from the changes in total risk, relative movement in the risk coordinate system, and other related factors, such as on-site inspections. Because there are an infinite number of combinations of probability and consequence that have the same risk, measuring just the change in total risk by itself is inadequate to assess a company's true potential for an event that affects its risk profile, for instance a large consequence accident.

Because risk is a relative measure, it matters less where a company starts in the risk coordinate system than how its risk changes over time and in what specific direction the risk values move within the coordinate system. Figure 8-6 shows the coordinate system divided into four quadrants. These four sections signify directions of movement indicating a change in risk from any previous point rather than well-defined areas. Each direction has different risk assessment implications, and we will discuss movement in each of these quadrant's directions separately.

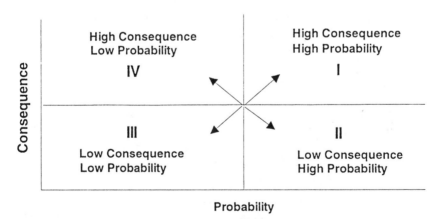

Figure 8-6. Risk movement direction categories.

[I] High Consequence, High Probability

A move in this direction signifies the worst possible case. To move in this direction, both consequence and probability had to increase. Hence the total risk value increased as well. Risk is also changing in such a way as to suggest the increased potential for events (accidents) with large consequences. Continued movement in this direction increases the chances of experiencing events that could financially affect the plant's ability to compete. The best label for this direction of movement is OOB—Out of Business!

How far can the risk value move in this direction? If consequence is being measured in terms of money, the upper limit could be related to financial assets of the facility. In general, the upper consequence limit is how much you can possibly ever lose from the occurrence of the event or events that make up the total risk value. Limits on the probability and consequence axes are discussed in the upcoming discussion of insurance applications of the coordinate system.

[II] Low Consequence, High Probability

This quadrant is labeled the annoyance or nuisance quadrant. Even though the probability of failure has increased, the associated consequence has decreased and total risk has been reduced. Movement in this direction is the second best direction. Risk is being reduced and so are the consequences. Even though failure frequencies are increasing, they are contributing less and less to the overall risk. The disadvantages of moving in this direction have more to do with failure frequency tolerance by people rather than risk contribution effects. Limits to movement in this quadrant will be discussed in the next section.

[III] Low Consequence, Low Probability

Movement in this direction is the objective of risk-based continuous improvement. It is not an ideal, but a realistic, achievable goal, with the proper measurements and management structure in place. Total risk is reduced and *both* the probability and consequence factors of risk are reduced from their previous values. From an assessment point of view, companies or plants moving in this direction are the least likely to experience large catastrophic accidents. They are the safest and most productive, the best of the best.

[IV] High Consequence, Low Probability

This region shares some general characteristics of region [II] - Low Consequence, High Probability, because it is possible to achieve the same value of risk from either combination of probability and consequence.

There are considerable practical differences, however. In region [II], the high-probability or high-frequency events supply ample data to use statistical methods to quantify measurement parameters and other aspects of risk measurement. Much is known about the behavior of the events because there are plenty of them to fuel the statistical tools that can supply descriptive information about the process. Because the consequences are relatively small and also prevalent, sufficient data for risk calculations are available. However, as you move in the direction [IV], less and less data are available to perform these calculations and the errors on the risk calculations become larger. The conclusion is that even though two companies can have the same total risk from either moving in direction [IV] or [II], the numerical uncertainties or statistical errors associated with these values will, in general, be different. The high-frequency, low-consequence region [II] events are managed at the plant level. However, the low-frequency, high-frequency region [IV] events, particularly these consequence events with costs above a company's deductible, are what keep underwriters and corporate risk managers awake at nights.

Aside from the numerical errors, there is another reason why movement in direction [IV] is different from moving in direction [II]. A company or plant that is experiencing events with increasing consequences at frequencies that result in changes of direction [IV] is assessed as more likely to experience larger, more catastrophic events resulting in severe consequences than a company moving in direction [II]. Major accidents in industry are seldom caused by just one isolated failure. They are almost always the final, cumulative result of a series of failures, incidents, and operating practices. A company whose total risk is reduced by moving in the direction of high consequence, low probability is heading in a direction where statistics are useless and the corporation's existence is at stake. One extremely high-consequence accident can change or destroy a company's financial future. In region [IV], statistics become useless. There are insufficient events to fuel conventional statistical engines. Without these tools, we enter a region where we are not capable of directly modeling the past to predict the future. There are little to no data on the frequency of major, catastrophic events inside a corporation. The risk measurement methodology discussed here is one of a series of tools that can be used to help people intervene in the large accident chain reaction, before the scenario fully and tragically develops.

To summarize this discussion, the quadrants are ranked from most desired to least desired below in Table 8-4. This is a risk assessment ranking, not a risk measurement ranking. The last column in the table indicates the type of risk measurement required to move in each quadrant.

Table 8-4
Risk Assessments Ranking of Movement in Each Direction

Quadrant	Risk Assessment	Risk Measurement
[III] Low Consequence Low Probability	Most Desired	Reduction
[II] Low Consequence High Probability	Satisfactory	Reduction or Increase
[IV] High Consequence Low Probability	Poor	Reduction or Increase
[I] High Consequence High Probability	Dangerous	Increase

Risk Assessment Example

Let's discuss risk measurement and risk assessment with an example. Suppose two companies, A and B, measure the risk associated with one event, and they plot their results in the risk coordinate system. It turns out that they both compute the same [p,c] coordinates, producing a total risk value of 25. The companies have the same total risk and are assessed as having the same accident potential. Time goes by, and next year companies A and B perform the risk calculations again. They each calculate their total risk as 10. This corresponds to a reduction of 15, i.e., from 25 in year #1 to 10 in year #2. However, they now are at different points in the risk coordinate system. From a total risk perspective, the companies enjoy the same risk reduction, however, in the risk coordinate system, their risk assessments are very different. See Figure 8-7.

Company B has reduced its total risk by moving in direction [II]. Although they have more frequent events, the overall consequences are significantly lower. This is not optimum, but the reduction is an improvement. In this case, plant management is more of a concern than risk assessment, as the company headed in this direction will eventually reach their "annoyance limit." This fact, however, would probably not be of a concern to the risk manager.

Company A definitely has the risk manager's attention. The movement in direction [IV] signifies an increased potential for higher consequence accidents. This result should be investigated to determine exactly why risk

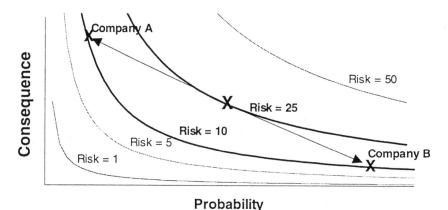

Probability

Figure 8-7. Properties of risk modification.

has changed in that direction. One common explanation is related to the measurement process. Remember Measurement Law # 5: "The human is an integral part of the measurement process." Because risk often is measured in dollars, a quick way to achieve the risk reduction goal (and possibly make the bonus) is to reduce costs in a way that affects the risk calculations without changing anything else. As a result, failures are less frequent, but larger consequences are experienced when something does fail. Reducing failures is fine as long as consequences are also reduced. A quality risk reduction is one that is obtained by reducing frequency of events *and* their associated consequences.

Insurance Applications of the Risk Coordinate System

Any time a graph is made, there is a tendency to look at the rigid axes, points, and lines in a clinical manner separate from reality. In essence, this is one purpose of making graphs—to isolate relationships from a process, so the process can be clearly illustrated. Representing total risk and risk modification in a risk coordinate system has the additional disadvantage of isolating relationships that are generally perceived as theoretical to begin with. Risk analysis is not a subject that gets much discussion at cocktail parties! The risk coordinate system does, however, serve as a useful tool to measure risk modification and risk assessment. It incorporates historical data along with human-based information, such as inspections, to assess the risk behavior of a system, plant, or corporation.

Reality is factored into the analysis by its very nature. The risk coordinate system structure is based on actual facility operation and experience. Every facility or corporation has limits on the maximum amount of any loss they can afford and on the minimum loss for which they require financial protection, generally in the form of insurance. There is an old saying: "Never insure something you can afford to lose." You may insure a car, a house, your life, and perhaps some irreplaceable antiques, but you probably do not insure a pen and pencil set, wheelbarrow, or ladder. Most people decide they can afford to lose items like these. In business applications, as you might expect, it is not that simple, but the same philosophy applies.

In this section, limits to consequence and probability are discussed that incorporate a facility's decision on how much they can afford to lose and how insurance protection fits into the coordinate system. The limits are actually thresholds that when reached, trigger certain ramifications.

Let's start with the consequence scale and assume the units are dollars. The particular names used for the limits will vary from company to company, but they all share common meanings [18].

The lowest threshold is called the *normal loss expectancy* or NLE. Normal loss expectancy is used for the most common events that usually involve minor repairs, replacement, and minimum business inconvenience. The dollar value is generally related to the insurance deductible. It represents a threshold that signifies the limits of the meaning of "minor" such as minor failures or minor business interruption. Below this level, the insured assumes all expense, similar to the way individuals do for some medical care and car repair. Below a certain threshold of expense, the individual pays all of the bill. Above that amount, the insurer starts to pay.

The second threshold is called *probable maximum loss,* or PML, and relates to the value of possible major accidents or failures requiring extensive repairs or replacement costs. These events do happen occasionally, especially from an insurer's perspective, which assumes the risk exposure for similar PMLs from many insureds. The cost implications of these particular events are fundamental to insurance pricing. For these events, those who experience the losses receive financial reimbursement. The insurer must collect sufficiently high premiums to have enough money on hand so that when a PML occurs, it can pay the bill and still be financially solvent.

The last level of consequence is associated with the maximum loss that ever can be achieved, called the *maximum possible loss* or MPL. This event is the worst case scenario, the catastrophic event that signifies the largest possible replacement costs. This value, along with the PML, helps define the insurance policy limits.

In practice, the art of underwriting involves negotiation with companies, and considers tradeoffs and other information to develop a sound, complete, insurance protection portfolio agreeable to everyone involved. The purpose of this discussion is to present the three consequence thresholds in a simplistic but genuine fashion, illustrating the real limitations that exist on the consequence axis of the risk coordinate system. Figure 8-8 shows a possible set of consequence axis limits.

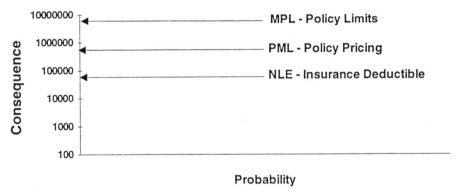

Probability

Figure 8-8. Example of company-specific limits of consequence.

Limitations on the probability axis are not as well defined. There are limits, but they are not clear cut, rigid, boundaries as with the consequence axis. The boundary is related to the capacity of the plant, corporate culture, regulatory agencies, environmental and consumer groups or the customer to tolerate increasing failures even though the consequences are reduced.

As discussed in the last section, movement to the right and down in the risk coordinate system is labeled as annoyance or nuisance. These words, annoyance and nuisance, were chosen for a particular reason. In general, a fly buzzing around your head could be considered a nuisance. It is relatively harmless, and you can continue to read or work without any real problem. If the fly increases its speed or others join in, you will get to the point that annoyance becomes intolerable and you take corrective action. The point is that there is a limit to how far risk can move in this direction. The limit depends on the event or process under study, but in every case, a practical barrier does exist. Whether it is the public's patience with environmental releases, the union's collective concern over safety, or the plant manager's judgment, the limit exists. The result is that movement in the nuisance/annoyance direction is fine for a while, but recognize that eventually a limit will be reached, perhaps unexpectedly.

The limits of the coordinate system are combined in Figure 8-9.

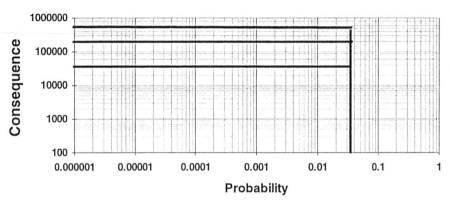

Figure 8-9. Example of company-specific limits of risk coordinate system.

Other examples of the coordinate system limits have been discussed in the literature [19]. In Figure 8-9 the coordinate system limits form the right and top sides of a rectangle. There are other suggestions to indicate that the limits form an arc sweeping from the consequence axis around to the probability axis. The particular configuration is not as important as the fact that limits exist. In essence, there are limits on how far you can go in terms of failures and consequences. However, there are no real limits on how much you can succeed or how much you can reduce risk. You might say there is a limit, zero! A zero risk facility or system is an ideal.

Let's spend a little time talking about time, that is, time between occurrences of postulated events such as the MPL or even PML. There is a practical limit how small the probability of occurrence can get. If we assume the universe is between 13 to 20 billion years old and if such an event has occurred once in this time (literally once in all time) then the average event frequency is 5×10^{-11}/yr. Clearly, if you see any failure probabilities around this value, they deserve a reevaluation. On the other hand, if the universe was created 6,000 years ago, as per Bishop Usher's Biblical chronology with no gaps, the failure frequency changes to 2×10^{-4}/yr

[20]. This figure does not seem as distant. In fact, many component failure frequencies are routinely calculated on the order of 10^{-6}/yr. The point is, any calculation of very low frequency events should be carefully reviewed for consistency, and more fundamentally, to see if the mathematical model makes any real sense.

Risk can always be reduced. You might respond to this by saying that the best that can be done is to reduce failures to zero. That is true, but from a risk measurement and assessment perspective, you can continue to prevent failures, which still translates into a lower probability and hence lower risk. For example, suppose you had two failures of a certain system last year. We will use frequency in this example because it is easier to relate to than probability. The failure frequency is 2 per year. The next year, no failures are observed. The failure frequency reduces to 2 failures in 2 years or 1 per year. The next year, by maintaining or improving the operations and maintenance, no failures are observed again. The failure frequency is reduced now to 2 failures in 3 years or 0.67 per yr.

Consequence reduction does not have the same characteristics as probability so there is no analogous example. Reduction in this area is more from improved operational and maintenance procedures. Engineering redesigns and systems analysis can also be factors in failure consequence reductions.

Risk reduction and continuous improvement go together. One cannot be accomplished without the other. When management decides to start continuous improvement programs, it is usually easy to "take the cream off the top" of a process. As time goes by and improvements are made, additional improvements become more and more difficult and perhaps expensive. Most processes have almost natural limits as to how much they can be improved without major reengineering. With risk as a measure, improvement (risk reduction) can be realized by maintaining low or zero failure frequencies for longer and longer time periods. System, equipment, and component reliability is always changing. If you can manage to go a year without failures and do it again two years in a row, you are reducing risk. You are doing this by matching the time-based system reliability requirements with the available resources. What could be better? Risk-based management is *calculated, not rash,* decision-making.

References

1. "Risk in a Complex Society"-Report of a Public Opinion Poll conducted by L. Harris for the Marsh McClennan Company, New York, 1980.

2. *Webster's Seventh New Collegiate Dictionary,* G. & C. Merriam Company, Springfield, Massachusetts, 1965, p. 743.

3. *Webster's New World Dictionary of the American Language,* Warner Books, New York, NY , 1984, p. 516.

4. *The American Heritage Dictionary,* Houghton Mifflin Company, Boston, Massachusetts, 1982, p. 1065.

5. Slovic, P.,"Perception of Risk," *Science,* 236, 17, April 1987, p. 282.

6. Starr, C., *Science,* 165, 1232, 1969.

7. Lewis, H. W., *Technological Risk,* W.W. Norton & Company, 1990.

8. Slovic, P., Fischhoff, B., and Lichenstein, S., "Facts and Fears: Understanding Perceived Risk," in *Societal Risk Assessment—How Safe Is Safe Enough?,* edited by Richard C. Schwing and Walter A. Albers, Plenum Press, New York, 1980, pp. 190–191.

9. Slovic, P., "Perception of Risk," *Science,* 236, 17, April 1987, p. 281.

10. Nisbett, R. and Ross, L., *Human Inference: Strategies and Shortcomings of Social Judgment,* Prentice-Hall, Englewood Cliffs, NJ, 1980.

11. Tvesky, A. and Kahneman, D., *Science,* 211, 453, 1981.

12. Cox, D. R. and Hinkley, D. W., *Theoretical Statistics,* Chapman and Hall, New York, 1974, p. 53.

13. Giuntini, M. E. and R. E., "Development of a Weibull Posterior Distribution by Combining a Weibull Prior with an Actual Failure Distribution Using Bayesian Inference," Fourth Space Logistics Symposium, Cocoa Beach, Florida, November 1992.

14. Giuntini, R. E., and Giuntini, M. E., "Simulating Weibull Posterior Using Bayes Inference," ARMS, 1993, pp. 48–55.

15. Atwood, C. L., "Correct Bayesian and Frequenctist Intervals are Similar," *Nuclear Engineering Design,* 93:145-151, 1986.

16. Millstone Unit 1 Probabilistic Safety Study, Northeast Utilities Service Company, July 1985, pp. 3.1-8–3.1-10.

17. Jones, Jaime Lynn, private communication, October 1993.

18. Rodda, W. H., Trieschmann, H. B. A., "Commercial Property Risk Management and Insurance, Volume I," American Institute for Property and Liability Underwriters, Malvern, PA, 1980.

19. "Risk-Based Inspection—Development of Guidelines," NUREG/GR-0005, Feb. 1992, pp. 25–30.

20. Minton, L. A. and Johnson, R. W., "Fault Tree Faults," International Conference on Hazard Identification and Risk Analysis, Human Factors, and Human Reliability in Process Safety," January 1992, copyright 1992 AICE, p. 67.

CHAPTER 9

Functional Risk Measurement: Incorporating Risk Into RCM

All business proceeds on beliefs, on judgments of
probabilities, and not on certainties.

Charles William Eliot

In this chapter, we will apply the concept of risk to reliability-centered maintenance. The result, called *Risk-Centered Maintenance,* or Risk-CM for short, adds the dimension of quantified risk to the functional emphasis of the RCM method. Before we begin a detailed description of this subject, let's talk about the reasons for explicitly incorporating risk with the RCM method.

Introduction to Risk-Centered Maintenance

As discussed in Chapter 1, reliability-centered maintenance began in the U.S. commercial aviation industry. Because of the physically compact, specialized, highly redundant nature of aircraft systems, the risks associated with failures were easily divided into four criticality classes: flight safety, operation, economics, and hidden failures. For the most part, these were, and still are, the categories necessary for developing a safe, effective, economical maintenance plan. The categorization of failures in the RCM process directly corresponded to a crude risk ranking, making the application of risk fairly straightforward for large passenger aircraft.

As RCM was applied outside of aviation, primarily in the nuclear power industry, the original four criticality classes continued to work well. This was because the range of systems to which RCM was applied included only the most vital areas, and omitted systems where the chance of failure was either low or the results were not severe. In these cases, the four categories used for aviation were sufficient for classifying the most important failure modes and assigning appropriate maintenance tasks.

Risk-centered maintenance was developed mainly because of the needs and problems of the types of systems to which it is being applied. As RCM is used for systems in many industries, one common characteristic is arising; the range of probabilities and consequences is becoming larger. This is generally either because the industries have multiple missions, or because their systems are more spread out than those of earlier RCM users, or both. It is no longer practical in many cases to choose systems for RCM analysis based upon *subjective* risk importance. The primary systems of such businesses as paper mills, refineries, natural gas, mining operations, and chemical plants are not as obvious as those in aircraft or nuclear power plants. Risk-centered maintenance uses the identical functional description of systems, subsystems, functional failures, and failure modes that RCM employs. It is different from the RCM method in that the criticality class is replaced with an explicit risk calculation. Using a quantitative value of risk instead of a coarse assignment (criticality class) allows a more complete description of the actual hazards that exist in these facilities. This chapter discusses how risk-centered maintenance compares to the standard RCM method and how to compute risk for failure modes from a practical and economic perspective.

Let's begin by reviewing how reliability-centered maintenance assigns an importance to failure modes. Each failure mode is analyzed via a series of yes or no questions that compose a decision logic tree. Each sequence of answers ends with a different action and a different risk assignment. The risk designations, or categories called "criticality classes," may vary in name, but generally relate to safety, production, economics, and hidden failures. Hidden failures are sometimes further analyzed with another specialized decision tree, refining risk assignments for this unique failure mode category. In RCM, risk assignments are made through these decision logic trees and are coarse classifications. Once a failure mode is classified into a criticality class, there is no further discrimination or ordering internal of the category. The method itself does not rank failure modes inside classes. The failure modes that fall into each category (safety, operations, economics, hidden failures) are all considered of equal importance. In practice, however, when people are correlating maintenance tasks with failure modes, there is usually an ordering based upon the team's judgment of importance. The criticality class is meant to provide general information about either the importance of preventing the failure or to the nature of the failure itself. When the range of consequences is small, this simple categorization procedure is good enough.

The risk-centered maintenance method replaces the criticality class identification with two separate fields: probability and consequence. The product of these two input parameters, risk, becomes an indicator of each failure mode's importance to the overall risk of the system. The independent assessment of frequency and consequence for each failure mode, and resulting calculation of risk, provide the ranking mechanism which is the unique benefit of risk-centered maintenance. This explicit calculation of risk is one reason we devoted a complete chapter to risk. We needed a foundation that illustrated both the power and limitations of the concept of risk measurement and assessment before we could apply it to RCM.

To conclude this preliminary discussion, we compare the spreadsheet versions of reliability-centered and risk-centered maintenance in Exhibits 9-1a and 9-1b to highlight their primary differences. A portion of the one-speed coaster bicycle example is used. If you went through the bicycle example, you will recognize the power transfer subsystem spreadsheet.

Exhibit 9-1a
Reliability-Centered Version—Power Transfer

Index	Description	Criticality Class
10100	**Loss of operator force transfer to pedals**	
10101	1 pedal failure	B
10102	1 pedal bearing failure	B
10103	Both pedals fail	A
10104	Both pedal bearings fail	B
10105	Poor pedal surface	C
10200	**Loss of forward force to rear wheel**	
10201	Chain failure	B
10202	Primary gear failure	B
10203	Foreign object between chain & primary gear	B

With risk explicitly computed as a numeric value, failure modes can be individually ranked from high to low risk. This ordered list will provide a priority ranking for choosing maintenance tasks to mitigate the occurrence of the failures. The risk computation replaces the criticality class assignment. The benefit is the replacement of the coarse risk assignments by an explicit numerical ranking.

Exhibit 9-1b
Risk-Centered Version—Power Transfer

Index	Description	Frequency	Consequence	Risk
10100	**Loss of operator force transfer to pedals**			
10101	1 pedal failure	0.01	10	0.1
10102	1 pedal bearing failure	0.03	5	0.15
10103	Both pedals fail	0.0002	100	0.02
10104	Both pedal bearings fail	0.001	40	0.04
10105	Poor pedal surface	0.1	2	0.2
10200	**Loss of forward force to rear wheel**			
10201	Chain failure	0.005	4	0.002
10202	Primary gear failure	0.0001	25	0.0025
10203	Foreign object between chain & primary gear	0.01	2	0.02

There are practical, sufficiently accurate, and cost-effective procedures for determining the frequency and consequence values needed to compute risk. These values do not have to be absolute, but must be comparable to one another, with the same measurement scale applied for the assignment of each. Once the risks are calculated, failure modes will be ranked from highest to lowest risk, relative to each other, not to an absolute external standard. Perhaps someday, when risk analysis for specific industries is standardized, there will be absolute risk standards, but for now and the foreseeable future, there are very few. Thus, for the risk calculations associated with risk-centered maintenance, accuracy is not as important as *consistency* in assigning probability and consequence values.

Let's now turn our attention to the specifics of assigning consequence values. In general we need to assign a number that describes the relative value of consequence associated with each failure mode. On the surface this seems like a difficult and highly subjective task. It is subjective, but for a very good reason. Regardless of the method used, the people doing the work, making the judgments, and correlating maintenance tasks determine the quality of the results. Their decision-making responsibilities are enhanced in Risk-CM. Consequence values should be determined by the collective experience and judgment of the team and others who are knowledgeable about the failure modes. However, for consistency, the team has the final say as to what value should be accepted. The goal should be unanimous agreement on each value assigned by team members, but, consensus is acceptable in most cases.

There are two difficulties with implementing risk-centered maintenance in practice. The first is the need for the overall group to accept the notion of risk. People accept failure frequency because they explicitly experience the effects. The failure of a process-critical motor that shuts the plant down for a week will be remembered long after the problem is corrected. People also accept consequences because they encounter penalties like lost production, high repair costs, and fines. These, too, are remembered long after the problem that caused the event is corrected. Failure frequency and consequences are *real*. People directly experience the events that are used to calculate their numerical values. With risk, it is a different matter. Before performing a risk-centered maintenance study, everyone involved must understand what is meant by risk.

The second impediment to implementing risk-centered maintenance is the acquisition of data. People may feel they have an insufficient failure database to extract the required failure frequencies. Most likely, they'll be right. Databases are usually designed to record equipment failures, not failure modes for RCM-based functional failures. Also, there will be failure modes identified by the analysis that have not occurred or have happened only once or twice in the plant's history. Defining failure probabilities for these situations is a routine and standard part of the work. Even though the exact data and format may not be in the database, with a bit of investigation, failure information can be accurately determined.

If you can remove the misconceptions that 1) risk is a theoretical tool without practical foundation and 2) meaningful risk calculations cannot be performed in situations with sparse data, the hardest part of the project is over.

Portability of RCM and Risk-Centered Maintenance Results

Before we discuss an example, there is one more aspect that must be mentioned. That is the portability of RCM and Risk-CM results. They can be reapplied to different systems. A study performed for one major plant system can be used as a basis for similar units that have the same functions. The subsystem functional division, failure mode identification, and the importance or consequence factors are approximately the same. While some changes will be made, there is no need to recreate the functional description from scratch. It is easier to revise a spreadsheet full of data than create a new one! The probability of occurrence and consequence columns in the spreadsheet will have to be updated to ensure that the failure modes, probabilities, and consequences reflect the new site configuration, economic environment, and personnel perceptions. Training plant personnel to perform the analysis is among the other factors to be considered, but a large part of the work is done. This is because RCM and Risk-CM look at systems from a functional perspective. They are not completely plant and/or site specific, so they can be re-applied to other systems with the same functions.

Now let's turn our attention to an example of Risk-CM that illustrates the principles previously mentioned in this chapter. The example is an actual Risk-CM project that shows how some of the practical difficulties we've just mentioned were overcome. It should give you an idea of the labor and time requirements for a moderately complex system.

Example of Risk-CM: A Gas Compressor System

The gas compressor system consists of a steam-turbine-driven, two-stage compressor that processes 40 psi of unsaturated hydrocarbon rich gas.

Before, during, and after compression, liquid condensate is separated from the gas stream by passing the gas through screen-lined cylinders called knockout drums. The pressurized gas (450 psi) and liquid condensate are then passed to other areas in the refinery for further processing. Cooling of the gas at inter-stage and after-stage compression points is performed by water-based heat exchangers that are connected to a cooling tower.*

Figure 9-1 displays the system functional layout, along with major subsystem boundary definitions.

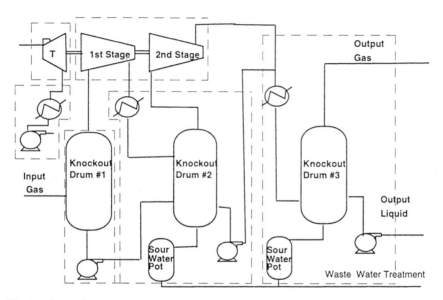

Figure 9-1. Gas compressor system with subsystem boundaries.

---

*Cooling will not be considered in this example because there are no aspects of Risk-CM that are unique to heat exchange. The cooling tower and the connected heat exchangers, known as the stream cooling subsystem, is an excellent example of a widely dispersed subsystem, however, as it contains miles of piping and covers a large distance. Heat input to the subsystem can occur at many locations around the facility and heat removal (output) at the cooling tower sites can be a considerable distance away. The location of cooling towers away from process areas is generally to ensure that cooling tower operation is not adversely affected by process failures and vice versa.

Index System

The index system is based on seven digit integers. It was discussed in detail in Chapter 4, however, because we use it here, we will review its syntax as it applies to the example. In general, the leftmost digit identifies the system. Because we are studying only one system, this digit is dropped for this problem. The first digit from the left now identifies the subsystem, the next two digits will denote the functional failure, and the last two digits the failure mode. For example,

10000: Subsystem #1
10200: Functional failure #2 of subsystem #1
10203: Failure mode #3 of functional failure #2 of subsystem #1

Part I: Failure Mode Risk Development

As with RCM, documentation provides a basis for discussion and reaching common ground about the system components and functions, provides a written record for future reference by personnel not involved in the study, and refreshes the minds of those who were. In this example, documentation was performed at a modest level, largely to help update people not involved with the initial work. A brief subsystem description and the information required for RCM or Risk-CM were recorded for each subsystem. This included equipment, special characteristics, in-interfaces, out-interfaces, functions, and functional failures. Exhibit 9-2 shows an example of the subsystem documentation.

Compressor Subsystem Definitions

From Figure 9-1, you can identify six separate subsystems. The seventh, Lube/Seal Oil, is a vital part of the system, but could not be easily depicted in the diagram. All seven subsystems are listed with their index values as follows:

Subsys#	Description
10000	Power Source
20000	Gas Compression
30000	Liquid/gas sep. (K.O.#1)
40000	Liquid/gas sep. (K.O.#2)
50000	Liquid/gas sep. (K.O.#3)
60000	Lube/Seal Oil
70000	Steam Condensing

Exhibit 9-2
Subsystem Documentation Format

Subsystem #10000   Subsystem Name: Power Source

Subsystem Characteristics: XYZ steam turbine (T12324), producing 4240 BHP. It is rated at 8900 RPM but actually rotates at a speed between 7400-7900 RPM.

Subsystem Special Considerations: Turbine is a multi-stage, condensing type for greater efficiency. It utilizes 400psi line steam and exhausts to 26–27″ Hg vacuum by condensing steam in a surface condenser.

In Interfaces:
  1) Steam at 380-400psi
  2) Filtered, cool oil
  3) Control

Out Interfaces:
  1) Steam at 26-27″ Hg
  2) 4240 HP at 7400-7900 RPM
  3) Hot oil
  4) Control

Functions:
  1) Produce required HP
  2) Maintain speed control of compressor
  3) Maintain steam/air boundary
  4) Maintain steam/oil boundary

Functional Failure #	Functional Failure Description
1	Total power loss
2	Partial power loss
3	Loss of steam/air boundary
4	Loss of steam/oil boundary
5	Loss of speed control

The analysis required approximately 1.5 labor months of effort. This is a relatively short time for analysis of a system this complex, largely attributable to available documentation and the cooperation of maintenance and operations refinery personnel.

Each subsystem was studied to identify the functional failures and failure modes that could cause each functional failure. Historical failure data were available to assist in this process. It is important to note that failure data only detail what *has* happened. The Risk-CM analysis also describes what *can* happen. Thus, all reasonable functional failures and failure modes were developed. The limit to the identification of potential failures was engineering judgment. The final risk ranking automatically selected and discarded failure events. This is one reason why failure database quality is not an obstacle. Failure modes describing what can occur and what hasn't yet occurred are a natural part of this work. The analysis uses whatever data values are available and accesses other resources for the other potential failure events.

Consequence Value Determination

Consequence values for the failure modes were assigned jointly by representatives from maintenance and operations areas. The group evaluated the failure modes, assigning a number of 1 through 4 to each, based on their experience. Next they repeated the process within each smaller group until the failure modes were completely prioritized. A final range from 1 to 100 was reached, where 1 indicated the least consequence and 100 indicated the greatest consequence.

The following categories or guidelines further describe values across subsystems and systems:

90–100	Safety related
89–60	Major effect on system function
	Long repair time
	High repair/replacement cost
59–30	Moderate effect on system function
	Medium repair time
	Medium repair/replacement cost
29–0	Small effect on system function
	Short repair time
	Small repair/replacement cost

The 0 to 100 range was expanded to 0 to 10,000 by squaring each value. The main reason this was done was to more accurately describe the relative consequence associated with the occurrence of each failure mode. These occurrences are intuitively non-linear. A change from 25 to 35 is nowhere near as important as a change from 85 to 95. The squaring operation models this perceived non-linearity in consequences. Now a move from 25 to 35 would be shown as a change from 625 to 1,225 (a difference of 600), and a move from 85 to 95 as a change from 7,225 to 9,025 (a difference of 1,800). It is important to note that the squaring process does not influence the order (or risk ranking) of the failure modes, but does change the contribution of each failure mode to the total plant risk.

Failure Mode Frequency Analysis

At the beginning of the study, there was some concern regarding the extent of data required to support such a mathematical study, as the failure data integrity varied considerably over the recorded time period. This is a common problem, and it did not adversely influence the project. The available data were analyzed for statistical trends and was used for frequency predictions. Additional failure data came from questioning maintenance and operations personnel who had many years of experience with the plant. The wealth of knowledge of personnel's experience was a primary factor for the success of the study.* Remember, in Risk-CM, consistency is more important than accuracy. Risk is used to rank the failure modes relative to each other.

Failure and repair data for all pumps, motors, turbines, and compressors in the gas compressor plant were extracted from the computerized maintenance management system and downloaded to a failure database. This failure database was analyzed for statistical trends for all components with 5 or more failures. For components with fewer than this number of failures, a *representative* (not necessarily mean) time between failures was computed. The trend procedure was discussed in Chapter 3. A brief review is given here before continuing with the example.

A trend is defined by three properties:

1. Trend Existence Probability: the probability that a trend exists among data values over the given time interval
2. Trend Type: improvement or deterioration
3. Trend Strength: how fast reliability is changing

Property #1: Trend Existence Probability

To assist in the decision process regarding the existence or nonexistence of trends, four independent test statistics are used. They are Laplace, MIL-HDBK-189, rank, and linear regression. The first two tests are

---

* Now, you do not ask maintenance or operations staff, who are genuine experts on the system, "What is probability of failure for the compressor steam turbine?" Collectively, they do know the answer to this question, but they will not generally understand what you mean. People can be used as sources of information most effectively and accurately if you know how to ask the questions and apply their responses. For example, if you wanted to know the probability of failure for the steam turbine, instead of even mentioning the word "probability," you could ask, "About how often have you seen the steam turbine fail? About once a year? Once every 3 years? Once every 10 years?" After querying many people you can put together all of the estimates and develop a failure frequency and error for the unit.

robust, nonlinear methods designed for small number of values. They also use the valuable information of the time from the last failure to the end of the time period. The last two tests use only the data values. The results of this analysis are displayed as a 4-tuplet, representing the percent probability that each test identifies a trend among the data values.

Property #2: Trend Type

Only improvement and deterioration trends are considered in this analysis. Techniques designed to recognize oscillating relationships require more data than are available in most failure databases. Oscillations can often be divided into a deterioration and improvement trend by segmenting the time interval.

Property #3: Trend Strength

A strength of the trend indicates how fast or slow reliability is changing. This measure is computed by comparing the mean time between failures (MTBF) and the predicted time to next failure (PTNF) computed from the observed trend analysis. If these two numbers are much different, then the observed trend is steep. A shallow trend is suggested when the two numbers are about the same. A completely flat trend occurs if the MTBF and the PTNF are the same.

Now back to the compressor system. Figure 9-2 shows an example from the analysis that displays actual statistical trend results in the failure data. The graph shows the time between failures measured in days on the vertical axis, plotted against the failure number on the horizontal axis. The last failure number, labeled by the double exclamation point, represents the future failure.

The dotted line shows the time behavior of the MTBF over the failure history. It is visual evidence to use in deciding if the MTBF is representative of future reliability. The straight line is a linear regression fit to the data points which are the large filled-in circles. The regression curve is a visual aid to show the overall behavior of the time between failures. The prediction of the straight line is indicated by the heavier line representing extrapolation into the future, i.e., the next failure. The connected crosses are the historical predictions for the next failure using all failures up to each point.

Trend analysis helps in the decision process but does not make conclusions. Remember, the objective of analyzing failure data is to determine a representative time to next failure (RTBF) to be used to denote the frequency of failure in a risk analysis. Thus, even if the trend statistics do not

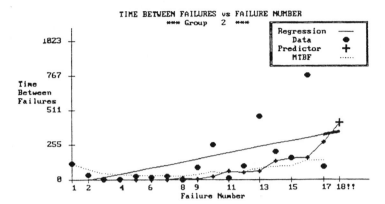

Figure 9-2. Improvement trend analysis example. (Reprinted with permission from *CheckMaint!*™)

give the existence of a trend a high probability, the team may still decide to use the trend-computed PTNF (Predicted Time to Next Failure) as the RTBF. This is what was done in this example. The team decided that the trend predictions were more representative of future failure occurrence than the historic MTBF. For the components that had too few failures for trend analysis application, the MTBF was used in the risk computations.

Computation of Risk

The computations were performed using common spreadsheet software. Table 9-1 shows the analysis for the first subsystem.

The RTBF (Representative Time Between Failures) column is printed in the spreadsheet because it is easier to read and more familiar to most people than probability. This number signifies the approximate time interval between experiencing failures. It is either computed from the trend analysis, from the MTBF, or compiled from personnel interviews. The failure frequency on a per day basis is calculated by taking the reciprocal of the RTBF and converting the units from per year to per day. This calculation will be done within the spreadsheet definitions, as part of the general risk formula in Equation 9-1. Because the risk values are going to be only compared to each other, the selection of time units does not affect the results. The "per day" unit is uniformly applied so that inter-system or inter-project risk comparisons can be made when other Risk-CM analyses are per-

Table 9-1
Risk-Centered Maintenance Analysis Spreadsheet Excerpt

Index	Description	Consequence	RTBF(yr.) ± Error		Risk ± Error
10100	**Total Power Loss**				**3.33 ± 0.07**
10101	Loss of lube oil pressure-unchecked	9025	61 ±	6	0.41 ± 0.08
10102	Loss of lube oil pressure-checked	625	12 ±	2	0.14 ± 0.05
10103	Blade failure-major	4900	14 ±	2	0.96 ± 0.28
10104	Radial bearing failure	2500	14 ±	2	0.49 ± 0.14
10105	Axial thrust-axial thrust brg failure	900	19 ±	2	0.13 ± 0.03
10106	Unstable rotor/shaft	400	19 ±	2	0.06 ± 0.01
10107	High lube oil temp	400	12 ±	1	0.09 ± 0.02
10108	Bowed rotor	900	29 ±	3	0.09 ± 0.02
10109	Turbine/compressor mis-alignment	900	12 ±	1	0.21 ± 0.03
10110	Coupling failure-diaphragm	9025	34 ±	3	0.73 ± 0.13
10111	Speed regulation valve	100	7 ±	1	0.04 ± 0.01
10200	**Partial Loss of Power**				**1.82 ± 0.59**
10201	Loss of blade efficiency - salt buildup	625	1 ±	0.3	1.71 ± 1.13
10202	Blade failure-minor	100	2 ±	0.5	0.11 ± 0.05
10300	**Loss of steam/air boundary**				**0.05 ± 0.02**
10301	Steam leaks in lines to atmosphere	100	5 ±	1	0.05 ± 0.02
10400	**Loss of steam/oil boundary**				**1.37 ± 0.82**
10401	Packing leaks-steam-oil contam.	900	2 ±	0.5	1.37 ± 0.82
10500	**Loss of speed control**				**0.76 ± 0.07**
10501	Governor probe failure	900	12 ±	2	0.21 ± 0.07
10502	Governor controller failure	900	12 ±	2	0.21 ± 0.07
10503	Actuator failure	900	12 ±	2	0.21 ± 0.07
10504	Steam inlet valve failure	625	12 ±	2	0.14 ± 0.05

formed. In mathematical terms, the risk value for each failure mode is computed from the consequence and RTBF column through Equation 9-1:

$$\text{Risk} = \{\text{Consequence}\} * \left\{ \frac{1}{\text{RTBF} * 365} \right\} \tag{9-1}$$

The error in the risk value is computed by carrying the uncertainty in the RTBF through the risk equation. It is, of course, good practice to include some level of error analysis in the Risk-CM study. The main reason is to ensure that risk estimates are reasonably consistent and to make sure that the errors do not exceed the risk values themselves. Because failure mode risks are developed to be compared to each other, error analyses offer no substantial improvement over the primary results. In addition, they are potentially time consuming. However, their relative values are important. For situations where absolute risk values are desired, the error analysis offers more of a pay back. Do not disregard the computation of risk errors. Just remember that you cannot get four-digit accuracy out of one digit estimates. Use the error analysis as a practical guide to ensure consistency in the risk calculations.

The transfer of errors through the risk equation above can be done in more than one way. The expression shown in Equation 9-2 is used to define the risk error for this example. Keep in mind, there are other ways to compute risk errors and you should explore which way makes the most sense for your application.

$$\frac{\text{Risk}}{\text{Error}} = \{\text{Consequence}\} * \left\{ \frac{1}{(\text{RTBF} - \Delta)} - \frac{1}{(\text{RTBF} + \Delta)} \right\} * \frac{1}{365} \tag{9-2}$$

where Δ is the error in the RTBF given in Table 9-1.

Using the spreadsheet functions, the failure modes only can be sorted and ranked in order from highest to lowest risk. A partial listing of the actual risk ranking of some failure modes for the overall gas compressor system is given in Table 9-2.

Table 9-2
Failure Mode Risk Ranking

Failure Mode Index	Description	Risk
40201	Sour water leaks in lines/drum	22.2
30201	Gas leaks in lines/drum	14.8
40301	Gas leaks in lines/drum	14.8
40506	Line leaks	14.8
50101	Condensate leaks in lines/drum	14.8
50301	Gas leaks in lines/drum	14.8
50501	Gas leaks	14.8
20501	Split line leaks	9.3
20502	Seal leaks	9.3
30101	Liquid leaks in lines/drum	8.9
40101	Condensate leaks in lines/drum	8.9
30302	Level controller failure	8.3
50201	Sour water leaks in lines/drum	7.4
70104	High or low level in condenser	6.8
20104	Impeller failure	6.3

As you can see from Table 9-2, the major risk contributors came from several subsystems and functional failures. Each line of the risk ranking gives the risk contribution from each failure mode to the system's total risk. Failure modes are the level where maintenance and operational procedures can modify the overall risk. They are equipment-based failures or combination of failures that can be prevented or caused by the way people interact with the system through maintenance tasks or operational procedures. The risk ranking shows the importance of each equipment-based failure relative to the other failures in the system.

Up to now, this procedure has given a systematic, engineering and user value-based relative risk ranking of system failures. They have been calculated by combining each failure mode's functional importance (consequence) and frequency (probability). This is what risk is all about. Figure 9-3 displays the complete risk ranking and represents the main result from the Failure Mode Risk Measurement part of the study. It shows the distribution of risk plotted against the failure mode indices for this example.

The gas compressor system contained approximately 100 failure modes. The horizontal axis is the index denoting their functional location. The vertical axis is their risk value. Linear scaling is used because most of the risk values are adequately contained with the given scale. The modes located on the extreme right side are relatively small and rapidly

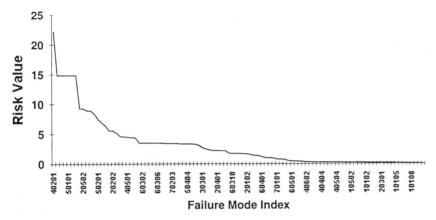

Figure 9-3. Failure mode risk ranking.

decrease in value. The total range of risk values is from 22 to 0.4, spanning approximately two orders of magnitudes. It is not uncommon to have a span of risk values of seven or more orders of magnitude. When constructing a functional risk analysis, you are considering the routine failures along with the possible, highly unlikely failures. Also, consequence values can further increase the range of risk values. In general, the range of risk values is determined by the scale used to compute consequences and failure frequencies or probabilities, and is specific to the system under study. The important point to remember is that just because a failure mode has a low risk, does not mean it cannot happen today or tomorrow. The maintenance analysis part of the Risk-CM method deals with practical ways to deal with this concern.

Part II: Maintenance Analysis

The next big step in Risk-CM is using the risk spreadsheet information to actually change the way maintenance is done. Using this information, you can address the group of failure modes that leads to your highest risk. Figure 9-4 shows the cumulative percent of total risk that is accrued as you move from highest risk failure mode to the right, incorporating each successive failure mode as listed on the horizontal axis. Notice that 80% of the total risk is contained within roughly 30% of the total number of failure modes. After performing a series of widely different risk studies, I've found that this relationship seems to be a general rule. This differs from the usual 80/20 rule, which says that 80% of the problems or activities comes from 20% of the total number of sources. Evidently, in Risk-CM, 80% of the total risk is contained within 30% of the risk contributors, which in this case are failure modes. Also shown in Figure 9-4 is a secondary rule, that roughly 60% of the total risk is contained within 20% of the risk contributors (failure modes).

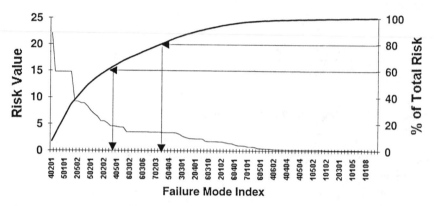

Figure 9-4. Cumulative % total risk and failure mode risk ranking.

The risk values associated with each of the functional failures can be computed by adding up the risks from their failure modes. Even though maintenance analysis cannot directly affect functional and subsystem level risk values, the calculations do show how risk is distributed among the system's functional parts. This information can be useful to the Risk-CM team in developing the maintenance design.

Figure 9-5 shows the functional failure risk distribution computed from the related failure modes. Notice that the failure mode index is in 100s to denote functional failures. The distribution of risk follows a more linear pattern which appears to be characteristic of graphs of functional failures. If failure modes are ranked from highest to lowest risk, the risk contribution of failure modes drops off in a highly nonlinear fashion. Figure 9-3 exhibited this typical behavior. However, when functional failure risks are plotted from highest to lowest, the successive risk contributions vary closer to a linear relationship. There are no established rules for how failure mode and functional failure risk contributions should vary.

The functional risks associated with each subsystem are summed to compute the risk contributions from each of the seven subsystems. The results

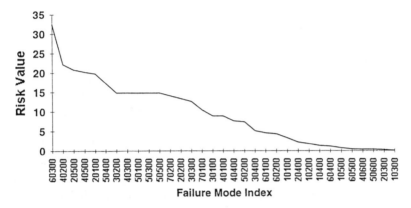

Figure 9-5. Functional failure risk ranking.

are shown in Figure 9-6. The subsystem risks are not ranked because there is only a small number of them to compare. Also, the ordering from left to right generally follows the process stream, so we can compare the risk associated with each subsystem in terms of its process location.

Inter-subsystem comparisons such as the one shown in Figure 9-6 have a valuable purpose. They provide the Risk-CM team with feedback about how risk is distributed among the subsystems. The subsystem risk ranking should agree with people's own perceptions. For instance, in this example, they should agree that Liquid/Gas Separation Subsystems #2 & #3 are the major risk contributors and the Power Source (the steam turbine) con-

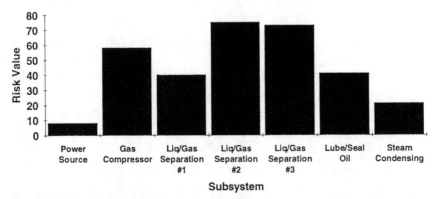

Figure 9-6. Subsystem risk contributions.

tributes the smallest subsystem risk. They should also agree with the complete ordering of subsystem risk. If they do not, or a consensus cannot be reached, then the cause of the discrepancy must be completely resolved. Why? The Risk-CM approach takes the large problem, System Maintenance Design, and divides it into a series of many small ones. In this case the large problem was divided into 100 failure modes. If people are going to believe the results at the failure mode level and therefore apply the risk ranking with confidence as a guide in developing a maintenance design, they must believe the results are an accurate representation of system risk. The subsystem risk calculation is a means for people to directly relate their own experience and judgment. If the macroscopic risk results are consistent with the team's view, then the application of the risk ranking will have achieved a degree of practical credibility. Because risk-based methods are generally new to the workplace, one seed of doubt about the relevance of the risk ranking can, and most probably will, multiply into a jungle of resistance. Most important, the experience and intuition of those intimately familiar with the systems must be acknowledged and generally can be trusted. Disagreements between results of the Risk-CM risk-ranking and the perceptions of those "in the know" may indicate errors or inconsistencies in the Risk-CM process. Inconsistencies should be resolved before proceeding.

The fundamental, detailed result of this study is the Risk-CM derived ranking of failure modes from highest to lowest risk. This listing forms a "standard" or "target" that provides a quantitative basis for measuring the effectiveness of the current PM/PdM program and information feedback on how maintenance resources should be allocated. It is a quantitative basis for identifying which failure modes should have the most attention and, just as important, which failure modes should have the least. It also helps in the decision processes that:

1. Determine where and when predictive technologies should be applied
2. Determine where and what employee training should be performed
3. Justify current or suggest new maintenance tasks
4. Justify current or suggest new maintenance task frequencies
5. Indicate areas of excessive maintenance or too little maintenance, and
6. Indicate engineering design changes that are needed

How the risk ranking is applied depends a lot on the nature of the company and/or facility. Here is one possible scenario. For each failure mode, the current PM/PdM tasks are analyzed for their potential to prevent the failures from occurring. Tasks may be added, changed, or deleted based upon their judged effectiveness to mitigate the high risk failure modes. Also, the ranked failure modes and corresponding task comparisons offer a basis to change current time-directed tasks (PM) to condition-directed (PdM), or to reassign task frequencies.

This procedure is systematically performed for all failure modes starting at the high risk end of the list. If tasks are judged effective, the yearly dollar cost associated with the necessary equipment and labor would be computed. Some failure modes may not be addressed currently by the PM/PdM program. For these, the cost estimates are zero. The cost column and the corresponding risk values are each summed. The two totals indicate how much of the plant's total risk is being addressed by the current PM/PdM program. The cost/risk comparison gives an engineering-based procedure for making decisions regarding the return on maintenance investment. Allocation of additional PM/PdM funds now can be related to how these funds will mitigate the expression of failure modes, and how much risk is going to be reduced by preventing these types of failures.

Earlier in this chapter, a comment was made that requires further discussion. The comment was that low risk failure modes still can happen tomorrow. This is true, but then how can you ever develop a meaningful maintenance design? The answer is that without taking the ideal, infinite resources to completely mitigate all failure modes, you will always assume some risk associated with system operation. Meaningful does not mean all-encompassing. We assume risks every day, so this is completely normal. The concern then becomes a matter of which compromise is the best.

Another general scenario for the application of risk-ranking is to divide the failure modes into two categories. The first category contains 80% of the system's total risk. For the approximately 30 failure modes in this problem, mitigating measures are taken that, in the judgment of the team, will adequately prevent their occurrence. For the remaining 70 modes where 20% of the system risk lies, a systematic sampling is taken and maintenance tasks are performed on varying schedules. If you completely ignore the failure modes that contain the lowest 20% of the risks, over some time period, failure modes will switch categories, i.e., some of the previously high-risk modes will be reduced and some of the low-risk modes where no maintenance is done will now be in the top 80%. Thus, the strategy is to completely treat the failure modes that contribute 80% of the system risk and to periodically perform maintenance on a controlled sampling of failure modes from the 20% group. This scenario is displayed in Figure 9-7.

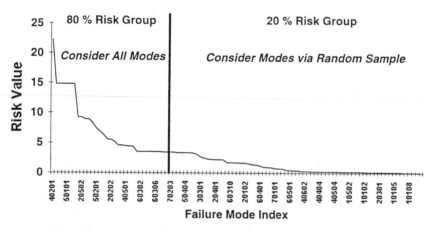

Figure 9-7. Maintenance strategy scenario.

As shown in Figure 9-7, the decisions as to if and how to assign maintenance tasks and frequencies lies completely with resources and the Risk-CM team. Probability, statistics, and engineering can assist in the decision process, but the complex analyses that weigh all of the variables, including the economics, are performed by the team members, not by computers. This fact is important because it reinforces a typical concern raised earlier about the amount and accuracy of failure data, and the difficulty in assigning consequence values. There is no reason to insist on *absolute* accuracy with a method that requires *relative* consistency and even then, still relies on people's expertise to make the final maintenance design decisions. In this sense, it is an imperfect solution method for an imperfect world. It is, however, difficult to think of a better approach to determine maintenance effectiveness or for designing a maintenance strategy.

Part III: System Risk Measurement

There are an infinite number of ways to document and monitor improvement program performance. The literature is filled with measurement strategies. They basically all translate to the lifeblood of industry,

$$$. Whether the program saves labor hours, reduces inventory stock, increases production, or decreases other costs such as energy requirements, the bottom-line is money. For RCM and Risk-CM programs, the objectives can be generally classified into three categories:

1. *Increase availability:* Incorporating reliability and production value of failures important for improvement. Actually from a viewpoint of plant performance, this *is* improvement.
2. *Decrease costs:* Availability improvement by itself only considers one side of the balance sheet equation. Obviously, if availability is increased but the associated costs are too high, the net economic result may be a decrease in the plant's bottom-line performance. On the other hand, one way to reduce costs is to downsize the workforce. This will appear as a windfall on the balance sheet until the plant systems respond with increased failures and increasing consequences. Thus, decreasing costs does not automatically imply staff reductions. Generally, any labor hour savings can be used to free highly trained plant staff to work on new projects that have the potential for increasing plant throughput, or for other saving measures such as decreasing contract labor or increasing product quality.
3. *Increase safety:* This goal is integral to every improvement program. While the actual work itself may not directly affect safety, we cannot separate human-based procedures from system performance.

Thus, the goal is to reduce costs and simultaneously increase plant performance. These goals may appear like "getting something for nothing," but this is not true. Plant systems are always changing with time. Consequently, even if maintenance and operations procedures were reviewed, or even if an RCM or Risk-CM study was performed a year ago, there are alterations that can be made to adapt procedures to the systems on a periodic basis. This process is represented by the term "Living RCM or Risk-CM Study." Using the Risk-CM version to explain the concept, it means that on a periodic basis, the frequency and consequences will be updated. The risk ranking will be redone and the ranked failure mode list reviewed and compared with the current operational and maintenance procedures. Changes, if any, will be made and new procedures or design changes performed. This process takes roughly one to two person-weeks of effort a year.

Now, let's turn our attention to a measurement strategy for Risk-CM. The goals we've just listed translate into one major objective: Reduce Plant Risk. Because risk is the primary decision variable in this version of

RCM, we can compute risk from the Risk-CM spreadsheet and observe how it changes in time. However, we must be careful that risk is reduced in the correct manner. This is the difference between "risk measurement" and "risk assessment." We will illustrate both of these activities for the gas compressor system.

Application of Risk Coordinate System

Risk Measurement

If I had to describe risk-centered maintenance in three words, I would say it is a *risk measurement tool*. The risk associated with each failure mode is defined by a point in a coordinate system. The horizontal coordinate is the failure mode frequency and the vertical coordinate is its associated consequence value. Thus, the risk distribution can be visually displayed as a scatter-plot by plotting the points from all of the system's failure modes. This is shown in Figure 9-8. The frequency and consequence axes use the logarithm scale to make plotting the complete range of values easier.

Total risk measurement can be done directly from the spreadsheet. We add individual risks from all of the failure modes together to produce the total risk from the compressor system. If we want to reduce total risk, we'll need this value to serve as a reference for future system performance. This performance will be the result of the team's decisions on the allocation of maintenance and operational resources.

From the scatter-plot, we can develop another way of representing the current risk associated with the compressor system. We can compute the "center" of the failure mode risk distribution. Because the mean statistic

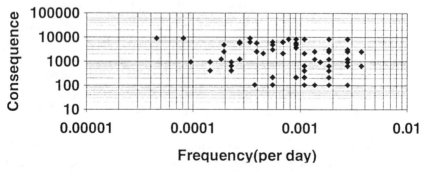

Figure 9-8. Compressor system risk state space representation.

is a way of representing the center of a distribution, we'll apply it to this situation. We use the arithmetic mean because we want to compute confidence intervals.

We define the scatter-plot center as the mean risk, R, as the coordinate location given by the mean frequency, f, and the mean consequence, c, each computed via the standard formulas.

The mean risk is a statistical center, but by itself represents an incomplete statement. We need to incorporate our degree of confidence with f and c into the coordinate system by computing the 95% confidence intervals. The confidence intervals are computed using the standard deviations from the frequency and consequence columns of the spreadsheet. In the coordinate system, these intervals define a box that marks the boundaries of our certainty of the mean risk value. Figure 9-9 shows the coordinate system with risk center box. The center of this box is the mean risk computed from the f and c values directly. The sides are defined by the confidence intervals.

We have performed all of the risk-based measurements that can be done initially. Now time must pass and at the conclusion of some time period, for example one year, a review of the consequences and failure frequencies are performed. Once this is done, changes in total risk can be measured by comparing the original total risk value computed before the time period to the result computed from the revised spreadsheet. If the decisions made during the Risk-CM project were correct, total risk will be reduced over time. Remember that plant risk behavior is a statistical process where we try to have as much control as possible.

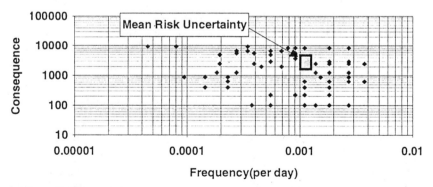

Figure 9-9. Compressor system risk distribution with 95% confidence limits on mean risk value.

Risk Assessment

We can use the compressor system to demonstrate the risk assessment utility of the coordinate system. After the one-year time period, we have a second spreadsheet containing updated frequencies and consequences. These values form another risk distribution in the coordinate system. Risk assessment involves determining the direction of the risk center movement and then, along with other factors, such as inspections, deciding if the facility is more or less likely to suffer high consequence events.

Using the four quadrants in Chapter 8, the most desired direction for the risk center to move is Quadrant #3 (towards the origin—the ideal of zero risk). This direction implies both failure frequency and consequence have been reduced. In general, movement upwards is bad (towards higher consequences) and movement to the right is bad (higher failure frequencies).

Let's discuss a scenario that illustrates some of these concepts. Figure 9-10 shows sample distributions with the original Risk-CM spreadsheet values and those from the spreadsheet as updated after one year. The total risk has been reduced. In fact, it has been decreased an impressive amount, from a year #1 total risk of 317, to a year #2 value of 79, over 400%. This was done by decreasing the frequency of failures by an order of magnitude. However, now the bad news, the mean value of the observed consequences has increased by a factor of two. The overall risk reduction was obtained by having fewer failures, but the failures that did occur had more severe consequences.

Clearly, risk reduction is good but the scenario indicates that more severe failures are occurring. What do these seemingly conflicting facts imply? Resolving this issue is what risk assessment is all about. It is probable that a system whose risk modification falls into Quadrant #4 (Up and Left) is more likely to suffer a catastrophic event than a system whose risk center has moved in Quadrant #3 (Down and Left). The risk assessment conclusion from the coordinate system results suggests that the current practices should be changed to reduce the trend of increased consequence failures.

All of the risk assessment calculations we've just been through can be done without the plotting shown here. When at all possible, however, I feel the graphs should be an integral part of the analysis. They give a visual picture of the risk distribution, risk reduction, and risk assessment that really cannot be achieved any other way. To most people, risk is an abstract quantity. The graphs provide something to which they can relate when risk and

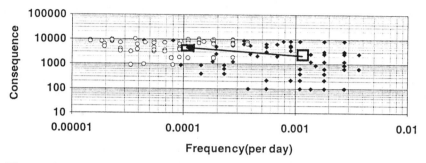

Figure 9-10. Two-year failure mode risk comparison.

its reduction are discussed. Risk is not something that can be measured directly. Risk must be computed from the observable event consequences and event frequencies. This is why a visual representation enables people to identify and begin to relate to risk more readily. With regard to presenting risk modification in terms of the coordinate system, the old saying of one picture is worth a thousand words is an understatement!

Common Maintenance and Operations Personnel Reactions

You might wonder how maintenance and operations staff members typically react to an RCM or Risk-CM type of review of procedures. As you might expect, reactions run the gamut from highly cooperative to . . ., well, let's say less than cooperative. For people on the negative side of the project, the most common reason for resistance, whether or not they express it directly, is their concern that the project is threatening their jobs. Labor savings is one well published result of these studies, although the published labor savings is only intended to indicate where less scheduled maintenance was required. The reports never used RCM or Risk-CM as tools to selectively eliminate jobs. The entire subject of labor force size is another issue beyond the scope of this book. More fundamentally, many people resist change, particularly to areas of their jobs where they are currently the "experts." A change means they may no longer be so familiar and competent. This is not only a threat to their jobs, but to their comfort level and esteem as well. Concerns such as these are among the reasons the RCM or Risk-CM team must communicate ongoing project plans and results to plant personnel. Communication skills should be considered when selecting workers for the team.

Generally, once the project methodology and results are communicated to plant personnel, reactions are positive, but with differing interpretations on how to use the results. Here's an example from a particular study. One of the dominant failure modes determined from the functional description was toxic gas-liquid leaks. Operations staff viewed these high-risk failure modes as safety related and assigned them a high-consequence value. They felt that the maintenance design should mitigate these hazards even though most of the leaks would not cause the plant to shutdown. The maintenance staff agreed that these failure modes were important from a safety perspective, but thought the high-consequence score was *too* high because most leaks did not cause immediate shutdown of the system. Maintenance personnel assigned high-consequence values to failure modes that caused the system to shut down, including modes that caused the compressor to fail.

The difference in perceptions between operations and maintenance was due to equating plant reliability to plant risk. Each group appeared to react to the project results based on how they were measured in their job performance. (Remember: You *are* what you measure.) Maintenance viewed failure modes with labor-intensive repairs as important (high-risk) and operations viewed the risk ranking of failure modes from the point of view of production or throughput loss. Because there are two factors involved in computing risk, focusing just on one factor can lead to conclusions inconsistent with the risk measure. High frequency *or* high consequence alone do not imply high risk.

These differences also highlight another fundamental characteristic of risk analysis. The consequence factor of risk does not distinguish between safety and reliability. Consequence values can be assigned or computed using many criteria. The important point is that events having like consequence values have like significance. All high-consequence failure modes represent serious ramifications to the plant regardless of whether they are major losses of production, high repair costs, large environmental fines, or worker injuries.

Successful RCM or Risk-CM projects are not produced by chance. They are earned by proper planning to lay a good project foundation and then lots of hard work. From a planning perspective, there are three key factors necessary for the success of RCM or Risk-CM:

- Employee and management understanding and acceptance of the risk measure
- Management commitment and leadership to support change
- Maintenance and operations teamwork and project ownership

All three ingredients are required for success. Take the time to do whatever is necessary to ensure you have a sufficient supply of each. The activities resulting from a successful Risk-CM study will undoubtedly save much more money than the study cost, but success is highly unlikely without a solid, immovable foundation built from the these three key factors.

CHAPTER 10

Operational Risk Measurement

When you cannot measure it, when you cannot express it in numbers, your knowledge is of a meager and unsatisfactory kind.

attributed to Lord Kelvin

Up to this point, we've focused on the large-scale, in-depth analysis and maintenance planning methods provided by RCM and Risk-CM. We've learned to investigate *all* areas of systems and their interactions so that as many contributors to risk as possible can be identified and managed. Unfortunately, time, money and people do not always lend themselves to so comprehensive a program. In these situations, operational risk management (ORM) provides an alternative approach that requires less time and fewer human resources.

Reliability or risk-centered maintenance are functional risk measurements that consider "what *can* happen," that is, they take an in-depth look at *all possible* failure modes and incorporate them *all* into maintenance planning. There is no question that such functional risk measurements are beneficial, especially for developing a complete maintenance plan and for managing low-frequency, high-consequence events. However, when there is a need for quick, low-cost reductions in risk, operational risk measurement is an excellent alternative. Operational risk measurement is performed only on process critical equipment. It involves a reorientation of data normally collected in plant operation and uses that data to identify areas of high risk and, just as important, areas of low risk. Its principles are applicable to the management of any business. A subset of RCM and Risk-CM, operation risk management looks only at actual losses, or what *has* happened and not at all possible losses. Its objective is to spend the available money and labor resources where the plant has historically lost the most, that is, the areas that have the highest risk. In addition, an effective operational risk measurement program can form the foundation for

RCM or Risk-CM projects as it points out the areas and systems that are candidates for further analysis.

Operational risk does not encompass all of the components of risk to a corporation. Not only do we consider only events that have already occurred, here we also limit our scope to identifying and quantifying only risk contributors that are within the plant boundaries. For example, companies that have plants on foreign soil in volatile areas of the world may find social instabilities a factor in risk management. With this method, however, we restrict ourselves to within the plant itself and deal with risk solely from a plant operational point of view, regardless of its location in the world.

Our discussions of computing operational risk measurement will include its data requirements. There are three fundamental characteristics that are common to all applications: (1) the time/date of failures, (2) the nature of the failures, and (3) the costs associated with the failures.

Accurate recording of the *time of failure* may appear to be an obvious and routine activity. Unfortunately, it is not always. In many cases, people will use the date the work repair order was opened as the failure occurrence time. Depending on what else is happening in the plant, this could be on the same day or could be several days later. The time of day information is usually not recorded at all. This information is very useful for analyzing the potential for subtle, synergistic effects, such as circadian influences, and is relatively simple to capture. Capture of the failure time is common with automated control systems like those involved in process control. For these systems, it is a simple matter of computer programming to record failure times.

The *nature of the failure* simply identifies what failed and how. A good way to uniformly and systematically record this information is with the use of failure codes. This minimizes the time required to record the event either in a computer management system or paper-oriented system. To some extent, it relieves the personnel demands in writing descriptions and also standardizes reporting responses. You want the codes to be as accurate as possible to capture the valuable failure cause or effect information. I use the word "accurate" here with deliberate intent. The personnel deciding on what code or codes to apply should be able to select the appropriate codes by scanning a list. You want to minimize the number of failures with the "miscellaneous" or "other" codes. I have seen up to 33% of all failures of one plant in the "misc" category. In this case, it meant that without interviewing the personnel directly who did the repairs, no information was available on the failure cause. Thus, from a practical point of view, no improvement could be made on 33% of the plant's failures! (This has been corrected since the observation was made.) Now work orders cannot be closed unless at least one failure code has been assigned. The failure code is a simple piece of information that is obvious to personnel at the time of repair. Unfortunately, it rapidly fades from memory as new problems continuously arise. That's why recording this information at the time when it is fresh in people's memory is invaluable for risk reduction. As an additional bonus, the use of standardized failure codes within a company is invaluable to computer-assisted analysis of failures and will eliminate almost all of the problems that have forced some RCM projects to resort to onerous paper-based analysis of failures.

The *cost of failure* is the most complex of the data items required for operational risk measurement. Cost cannot be readily entered at the time of the failure occurrence because it includes many factors. These range from tangible items, such as cost of parts and labor, to those which are less obvious, including lost production, insurance increases resulting from the failure, workers' compensation, and litigation costs. Cost of failure is the fundamental measure of operational risk.

Components of Operational Risk

Before we can measure operational risk, we must first identify its primary components, which are (1) equipment risk, (2) product risk, and (3) people risk (Figure 10-1). The risk contributors interact with each other in both direct and indirect pathways. Every plant has unique factors that compose its specific risk components.

Operational risk measurement begins with data gathering and summarizing. The data for this process are normally collected during plant operations. Here is an overview of the steps involved:

1. Look at the frequency and costs attributable to *equipment* failures. This involves gathering the numbers that tell you how money *left* the company to pay for costs incurred as a result of equipment failures and how often. Costs will include equipment replacement, purchase of parts, and both company and contract labor fees. These expenses should be accrued over a fixed time period, which is generally one year.

2. Gather information about the frequency and costs incurred because of *production* failures. This actually means summarizing information about how much money *didn't come into* the company because of production failures and how often these losses of income occurred.

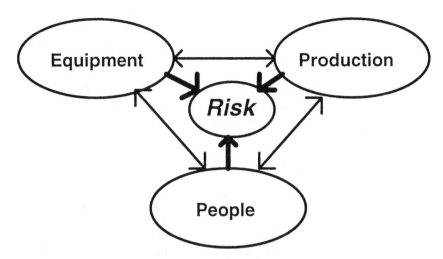

Figure 10-1. Major contributors to operational risk.

3. Tally the costs and frequencies of failures that are related to *people.* These include the costs and frequencies of injuries requiring workers' compensation and other payments such as injury-related liability. Less dramatic, but still costly is the loss of productivity that occurs with failures in the equipment and/or process. In most cases, employees dependent on the equipment/process must wait, unable to produce much of anything, until the failure is overcome. This is money spent without return.

4. Convert all of the costs and frequencies you have gathered into *risk,* generally expressed in terms of dollars.

5. Identify the "heavy hitters" or largest contributors to your list of risks. These are the areas that should be given immediate attention and resources.

Equipment, product, and people risk components are described in more detail in the next sections. Each component is described individually to give you a sense of what items fall under each category. The goal is to reorient your current data to give an *operational risk profile* of the plant. An operational risk profile is a "picture" of the effective dollar loss associated with each of the three primary contributors to risk: equipment, product, and people. It is multi-dimensional, attempting to view risk from key perspectives within each of the three contributors' realms. The aims of the operational risk profile are to find the most effective way to reduce risk by identifying specific areas that need improvement and their contributions to overall risk, and to prioritize the ways improvements can be made.

Where do you start in developing an operational risk profile? We'll go through a series of examples that will show some approaches. Let's begin with an in-depth look at the contributions equipment makes to risk.

Equipment Contribution to Operational Risk

The fact that equipment is a major factor contributing to operational risk is no surprise because maintenance tasks are performed on equipment. Equipment operating together in a harmonious fashion produces products. People interact with the process through equipment. The difference in developing a people vs. equipment risk profile is how the data are used more than the data items themselves.

Here's our strategy for developing the equipment contribution to the risk profile: we're going to take a "bottom-up" approach to assessing system risk. That means we'll first look at all of the failure effects that are logged in the data. Next, we'll find all of the equipment failures that are logged. We'll move up the hierarchy, grouping them by subsystem, and then grouping again by system to get a total picture of the operational risk due to equipment. Here are the details:

The most basic level of observable system failure is the *failure effect.* This is identified by the failure code discussed earlier. Ideally, the failure code would denote the root cause of failure. Root causes are not usually the obvious symptoms of failure and require considerable investigation. For example, suppose the observed failure effect is a leaking pump seal. Further analysis indicates that it was caused by motor-pump misalignment. Further analysis indicates that the misalignment was caused by improper pump installation. Is this the root cause? Why didn't the people who installed the unit do it correctly? The scenario could be continued. The point is that at some level of analysis, a line must be figuratively drawn and someone needs to say that *this* is the root cause. Thus, the term "failure effect" signifies the observable level that is a compromise between analysis detail and practically achievable field information. Keep in mind that the more detail placed in the failure effect level (for example, in the descriptive failure codes) the more accurately risk reduction actions can be applied.

The next level in the equipment hierarchy is the *type of equipment* that failed, such as transformer, compressor, motor, pump, etc. As we mentioned, the equipment analyzed in the risk profile is a subset of the plant's total equipment. It is the set of process-critical equipment that affects production and/or profitability. This is where the best return, that is the greatest risk reduction, is obtained for the investment. As mentioned in the "Computerized Maintenance Management Systems" section of Chapter 4, different levels of detail can be kept for equipment that differ in importance levels.

The next level of analysis is the *subsystem.* Here we must consider how equipment interacts with its functional environment. Its position in the process can affect equipment performance and failure history (and future). For example, considering a pumping subsystem. It is true that the pump actually "manufactures the product," in this case, transportation of a gas or liquid. However, a *combination* of equipment acts as one functional unit to effectively perform this function. This unit includes the electric switch gear, motor, pump, and pressure control valve. A failure of any one piece will cause the loss of the pump's primary function.

Another reason for grouping equipment into functional units is to identify potential synergistic interactions between equipment not recognized by standard reporting procedures. Failure frequencies and costs are routinely tracked and understood according to equipment type, but not necessarily by subsystem. It is not always clear how the subsystem is performing, because people interact with the subsystem from an equipment level. A switch gear, motor, pump, and pressure control valve may have each failed only once in a year with little or moderate associated costs. In relative operational terms, this performance may not be excessive. However, if the subsystem has failed four times that year, when you add the costs of the four separate failures together, the subsystem risk may be significant. This could indicate that a synergistic interrelationship among the failures exists and justifies further investigation.

The next layer of analysis is at the *system level.* Failure and cost data from all of the related subsystems are combined into sets denoting plant system operations. Another layer could be defined as a *super-system* or plant. The actual structure of the equipment hierarchy is clearly plant dependent. To give you an example, Table 10-1 displays some of the ways a pipeline company looked at their failure data. Each column of this chart itemizes the contributors to one perspective of equipment operational risk assessment. Subsystems are made up of many pieces of equipment from the Equipment Type column. A failure in any piece of equipment is a failure of the subsystem.

Table 10-1
Operational Risk Profile: Equipment Category Framework

Failure Effect	Equipment Type	Subsys	(System) Pipeline Segment	(Super-System) Pipeline System
Mechanical	Switch gear	a	BOS—NYC	Northeast
Electrical	Motor	b	NYC—PHL.	
Seals	Pump	c	PHL—DCA	
Bearings	PCV	d		
PCV		e		
Misc.				

The individual data items in Table 10-1 can be combined to produce risk results to be directly applied by plant personnel to improve plant performance. The risk results are designed to determine which failure effects, equipment, subsystems, and systems have the highest and lowest risk. This risk analysis only goes halfway. It identifies, quantifies, and ranks operational risks using an equipment category list such as shown in Table 10-1. The *real* challenge is for the experts on the plant's particular equipment and systems, the operations and maintenance management, to decide on the specific risk reduction actions. The beauty of this method lies in the simple fact that improvement through risk reduction actions and potential cost savings are based on facts. The operational data that fuels the risk analysis is actual history, not conjecture. It cannot be changed or disputed.

To give some examples on how to perform this kind of practical risk analysis, let's discuss the set of results shown in Figure 10-2.

The first plot shows a frequency distribution of the number of failures in each component-failure effect category. In theory, every failure effect could be related to every equipment type, but in practice there are usually specific failure effects and equipment types that always go together. This is a standard Pareto chart commonly used in industrial measurement. If all failures had the same consequences, that is, if they all had the same cost implications, the frequency plot would be both necessary and sufficient for decision-making. Because they obviously do not, the frequency plot is neither necessary nor sufficient.

The middle plot shows the dollar converted costs of these failures and depicts purely financial implications. It does not indicate if the costs are

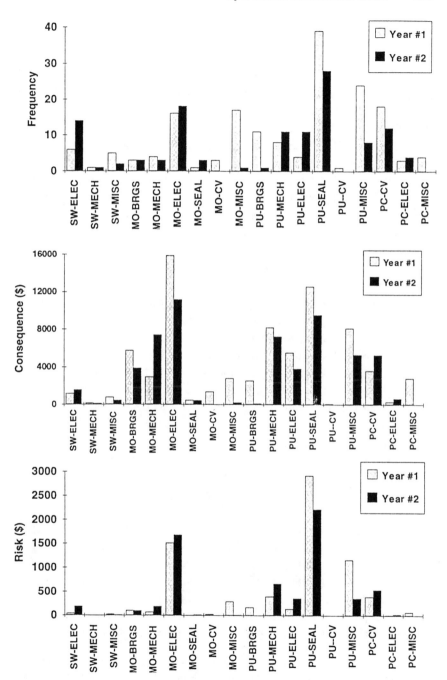

Figure 10-2. Risk-based dominant failure effects.

associated with recurring, frequent problems or one event mishaps. This information also falls short of adequate for decision-making.

The bottom plot shows the risk result. By multiplying the frequency of events in the top graph by their associated consequences in the middle plot, the risk results are computed. Notice that in this case, the risk-based dominant component-failure effects are very easy to spot. Cost and frequency-based dominant failure modes are somewhat less clear.

In practice, all three plots provide information that is helpful in planning where and how to concentrate resources for improvement. However, the more comfortable people become using risk-based methods, the more reliance is placed just on the risk results. In essence, risk analysis is akin to playing the odds with the deck stacked in your favor. By marrying frequency and consequence of events, you are best able to distinguish what is important and what is not.

The previous discussion displays operational risk by examining equipment failures and their associated costs from a functional perspective. The risk results are expressed by function and functional failure (failure effect). Another measurement of operational equipment risk can be performed by studying the distribution of repair times, which ideally includes the costs associated with repairs or equipment replacement.*

Next, we have to determine which types of failure events have the largest risk contributions. Failures are categorized by repair time intervals. The "types of failures" are the repair time intervals. For example, consider the frequency distribution shown by Figure 10-3. This frequency plot indicates that there were 28 failure events that had repair times between 0 to 5 labor hours. In the 6 to 10 labor-hour interval, there were 17 events. From a frequency perspective, the most important type of failures were those with the smallest labor repair times.

It is interesting and perhaps just human nature, that often the highest frequency types of repairs are considered the major problem areas. This is understandable for two reasons. First, people learn by repetition. Remembering is a type of learning. The more a task or set of tasks is performed, the more they are remembered. And, therefore, the more they are perceived as "major problem areas." Second, the people making the repairs actually experience the repair frequency but do not see the overall repair expense for labor to the company. These figures are seen by accounting

---

*If you cannot compile this information, use labor hours as a measure of total repair costs. High labor-hour repairs can be assumed to indicate high repair total costs. This is not perfect, but it is a safe assumption.

and management who are at least once removed from the actual repair activity. Consequently, the type of repairs labeled as problem areas are those performed most often by maintenance, and the most expensive are the ones identified by accounting and/or management.

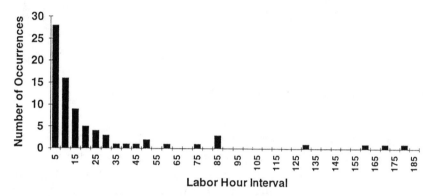

Figure 10-3. Labor hours frequency distribution.

Now to compute the risk associated with repairs in each labor-hour interval, we combine the value of frequency and the labor-hour interval. The value of frequency is divided by the total number of events (in this case 79) and converted to a probability of occurrence. The labor-hour interval is converted to consequence (dollars) by multiplying the interval time by the average labor-hour cost. For example, in Figure 10-3, the risk computation for the failure events in the 0–5 interval is as follows using an average labor rate of $30 per hour:

Risk [0–5] = 28/79 × (5 × $30/hr) = $53

When this is performed for each labor repair time interval, a very different distribution of risk is seen. In Figure 10-4, note that the highest risk contributors are events with labor repair times in the 80–85-hour category. In general, the high labor repair hour categories are seen to be dominant risk contributors.

How is this information used? The risk distribution shown above identifies and ranks the risk contribution from each repair time interval. The actual failure events that compose the high-risk intervals are reviewed by

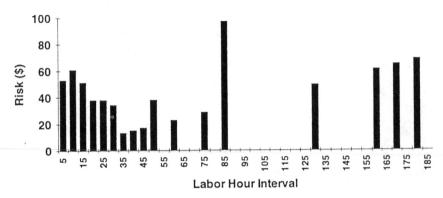

Figure 10-4. Labor-hours risk distribution.

an analysis team and/or management to decide how best to reduce the observed risk exposure in these areas. It is these areas, based on the recent past, that afford the highest risk reduction return or risk-benefit. Analysis of these events may or may not reveal any similarities. The main point is that the analysis team now has the clear challenge of devising an improvement strategy to reduce or eliminate these classes of failures knowing that if accomplished, the largest risk reduction for the dollar spent will be achieved.

It would be nice to have the method for how to remove failures and their associated costs displayed at the end of a computer printout. If plant management was that simple and logical, somebody would be marketing a computer program to do it by now, and many of us would have different careers.

You may have wondered why the risk analysis of labor repair times is necessary if you perform a comprehensive equipment functional analysis. The reason is simple. Permit me to start my answer with an analogy. If you wanted to view a complex three-dimensional object, you would look at it from at least three different directions. The more complex the object, the more directions from which you must view it to comprehend its shape. Plant operational risk is a very complex object. To comprehend its shape and structure, we apply different procedures that allow us to observe it from different perspectives. So far, we have viewed it from the equipment side using two different methods—failure mode and labor-hour analysis—to understand its structure. Next, we turn our attention to the viewpoint of production contributions to operational risk.

Production Contribution to Operational Risk

This major contributor to risk is composed primarily of the economic implications of lost production due to failures. Lost production is commonly interpreted as downtime or time lost due to equipment failures. Certainly equipment failures compose the major part of risk quantified under this component. However, there are other ways to lose production. Production rates can be reduced because of insufficient raw material supplies or bottlenecks in shipping, storage, or packaging. These types of losses are harder to compile because data values are usually not maintained on losses due to rate reductions. In discrete production facilities, changeover times are another source of production losses. Even though product setup and changeover are necessary, any time a line is down, money is being lost. There is an operational risk associated with these activities that cannot be reduced to zero, but clearly can be minimized. Any time a production line is not producing at its designed rate, it contributes to operational risk. This includes scheduled maintenance and turnarounds.

One of the most important types of lost production is not related to rate of production but to product quality. Lost production due to equipment failures or management problems are internal concerns not affecting customers. Product produced out of customer specifications, not produced on time, or not produced to company quality standards has the potential to damage customers' perceptions of business performance reliability. Quality is another dimension of operational risk. Loss of customer confidence is difficult to accurately measure, and it is even more difficult to regain once lost.

The analysis of operational risk from production discussed in this section follows a procedure similar to that we used in the risk analysis of labor hour associated with failures. We define lost production as the product of capacity and time lost. For example, if a production line has a designed rate of 2,000 lb of product per hour and it is down for 2½ hours, then the lost product is 5,000 lb. In this case, we measured lost production in terms of lost pounds.

If production systems are equipped with on-line spares, then a failure of the primary unit activates the backup unit and production continues, perhaps at a reduced rate, depending on the size of the standby equipment. The point is that production rate reductions are caused by equipment failures, but not all equipment failures affect production.

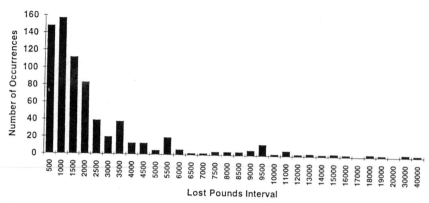

Figure 10-5. Lost pounds frequency distribution.

There is a direct connection between production losses and equipment performance. Thus, at least in theory, it is possible to link specific failure effects, equipment, subsystems, etc., to specific production losses. In my experience, companies routinely measure equipment failures and production losses separately. It is difficult with this type of separate accounting to compute the *total* cost of individual failures, i.e., labor hours, parts, contract labor, and lost production. This separatism is fostered by non-integrated information systems. Automated control systems that record production activities do not routinely communicate with maintenance management systems where equipment failures are recorded. This is one area that is being addressed by "reliability management" computer systems now in design. The connection can be made without major software modifications to current systems. The production control system must identify the equipment that has failed. Then on all work orders from the maintenance system that are written to repair the equipment, an index number is assigned. This index acts as a logical bridge that connects specific equipment repairs to a lost production event.

Because the production loss ↔ equipment failure link is not generally transparent, we analyze the production contributions to operational risk separately. Going back to the units established in our example above to measure lost production, we identify all lost pounds events and develop a frequency distribution showing the number of times a production failure occurred in each lost-pounds interval. Figure 10-5 displays a sample distribution for one year's worth of operational data.

We apply the same methodology as with the labor-hour interval risk analysis. Figure 10-5 shows how often failures yielded losses in each category. For this year, there were 719 lost-pound events, providing the 719 data points that together make up Figure 10-5. The most frequent loss category was in the 501–1,000-pound category. The least frequent category was the 20,001-30,000-pound interval that had no values for the year. From the graph you can compute that roughly 80% of the time, a production interruption event resulted in 4,000 or fewer lost pounds. The other 20% of the lost pound frequencies are shown by the remainder of the plot. This might tempt you to believe that the most concern should be placed in analyzing the particular events with lost pounds from 500 to 4,000. From a frequency standpoint this is true. But *frequency* is not the issue; the *net amount of profit lost* to the company is. This is how a company can combine historical loss magnitude and frequency to make the best business decisions for the future.

In essence, we want to use the experience gained from the past to improve the operational future. We have two sets of information to use along with the expertise of the plant's personnel: loss frequency (probability) and loss magnitude (consequence). These two stochastic processes *together,* not separately, determine the observed production performance. Because the two processes naturally interact, we can compute the interaction or risk distribution. For an example of the calculations that compute the risk distribution, let's take the lost-pound interval [0–500]. From Figure 10-5, we see that there were 148 out of a total of 719 events in this category. In this case, the cost of product was assigned as $0.60 per pound. The risk calculation for the [0–500] lost product interval is:

Risk [0–500] = (148/719) × (500 × 0.6) = $62 / year

Figure 10-6 shows the complete risk distribution computed from the data in Figure 10-5.

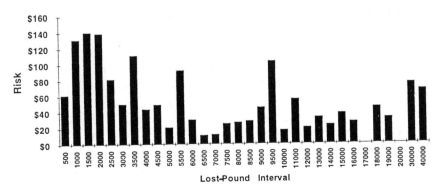

Figure 10-6. Lost production risk distribution.

The risk distribution shown in Figure 10-6 indicates that the major lost-pound intervals are 1,000 through 2,000. These results roughly agree with the frequency distribution, Figure 10-5. The interesting information in this plot is the set of lost-pound intervals that compose the next three risk contributions. Notice that the next three largest risk lost-pound intervals are 3,500, 9,500, and 5,500, respectively. *The interval with the highest frequency (500 pounds) is not a dominant risk contributor.* To observe the ranking of the risk associated from each lost-pound interval, the sorted risk values are given in Figure 10-7. This presentation format shows the relative risk ranking or importance of each lost-pound interval and helps decision-makers determine the specific category of events to analyze. It is interesting to note in this example taken from actual plant data, the smallest lost-pound interval (that had the highest frequency of occurrence) is ranked next to the largest lost-pound event category.

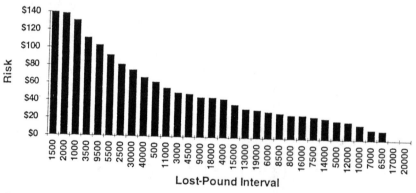

Figure 10-7. Risk ranking of lost-pound intervals.

In this example, the analysis team has its work cut out for them. Ideally, you want to have a risk distribution that drops off more steeply. That behavior would allow you to select the dominant risk contributors with more assurance that the risk-benefit of addressing additional lost-pound events is small. In this example, however, the risk contribution drops off at almost a linear rate. See Figure 10-8.

The cumulative percent risk is plotted as a function of the risk ranked lost-pound interval. In this case, seven lost-pound intervals contain 50% of the observed operational risk from production effects. Eighty percent (80%) of the total risk is contained in the first 16 intervals or 80% of the

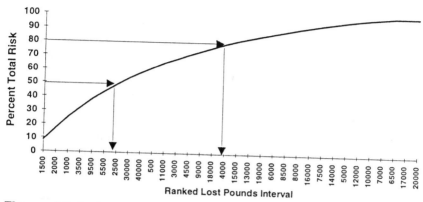

Figure 10-8. Cumulative risk contribution from ranked lost-pound intervals.

risk is contained within 50% of the lost-pound intervals. The more operational risk is concentrated in specific lost-pound intervals the easier it is for the analysis team to decide on which intervals to study and which to exclude. In this example, the risk-benefit decision is not simple. Risk is almost distributed linearly between lost-pound intervals, making the analysis stopping point more an economic decision than a technical one.

Operational Risk Assessment

The methods discussed in the chapter to this point have dealt with risk measurement. You can apply the risk assessment methods developed in Chapter 8 to the operational data compiled for the operational risk profile development. The reason the assessment methodology is an important part of risk analysis is because risk reduction is an incomplete description of performance. As in Chapter 8, it is possible for a company to enjoy a reduction in operational risk, yet from an assessment perspective be more likely to suffer high-consequence accidents and therefore be a poorer risk to an insurer. In hindsight, operational risk reduction is all that is required. Reducing operational risk while not suffering high-consequence events when it is a matter of history is great. The only trouble is that insurers and business managers are concerned with the future, not the past. This is why we need to assess *how* risk is reduced in order to predict the future risk as a company and as an insured.

Quantitative risk assessment requires operational data for at least two time periods. Using the current example, data for the previous year's lost production are used. The lost production data are categorized into lost-pound intervals using the same procedure previously discussed. The partial listing of the two years' lost-pound data is given in Table 10-2.

Table 10-2
Lost Production Summary Data by Lost-Pound Interval

Lost-Pound Interval	Number of Occurrences	
	Year #1	Year #2
500	38	148
1,000	67	157
1,500	24	112
2,000	37	83
2,500	29	39
3,000	20	20
3,500	11	13
4,500	19	13
5,000	5	5

Consequence terms are calculated by multiplying the lost-pound values by the previously used conversion factor of $0.60 per lost pound. The corresponding probabilities are computed by dividing the frequency values by the total number of occurrences each year. The resulting risk coordinate system representation of the two years' lost production is given in Figure 10-9.

As you can see by the dispersion of the individual points, it is hard to visualize any differences between the two years' risk performance. To quantify the difference in production-related risk, we compute the "center" of each distribution. The mean values of consequence and the mean values of probability were computed for each year's data. For this particular calculation, where the representation is in logarithmic coordinate system and the variables contain probabilities, we apply the geometric mean. Lost-pound intervals with 0 occurrences are excluded from the calculations. Before we plot the "risk centers" for each year, let's look at the numerical results given in Table 10-3.

These results are all in the right direction. Every quantity has been reduced from year #1 to year #2. Risk reduction does not always imply that *both* consequence and probability are reduced. In this case, the results indicate the plant enjoys both a good risk reduction and a good assessment of future risk performance. The "risk centers" are plotted in Figure 10-10.

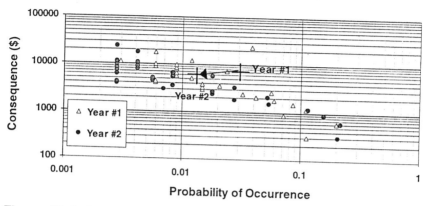

Figure 10-9. Risk coordinate system representation of lost production.

Table 10-3
Risk Center Results

Quantity	Year #1	Year #2	% Change
Avg. Consequence	$4,029	$4,007	−0.5%
Avg. Probability	0.019	0.011	−42%
Avg. Risk	77	43	−44%

The assessment quadrants labels, discussed in Chapter 8, are also given to identify the risk assessment result.

The risk center movement shown in Figure 10-10 indicates the company has obtained very close to an optimum risk reduction. Quadrant [III] is the most desired type of risk assessment result. The probability (or frequency) of down-time events has been reduced and the consequence of the events has also been reduced. The optimum reduction occurs when the risk center moves directly towards the origin, that is, towards the ideal of zero risk. In practice, any motion into quadrant [III] is good.

Before we leave the subject of risk assessment, I want to discuss the practical utility of the graphical method. First of all, the risk assessment information could be obtained without the aid of any graphics. The risk centers' calculations are simple algebraic operations and the direction of motion can be obtained through simple trigonometry or in numerous other ways. The reason the procedure relies heavily on graphics is to enable people who might not be current in applied statistics and trigonometry to

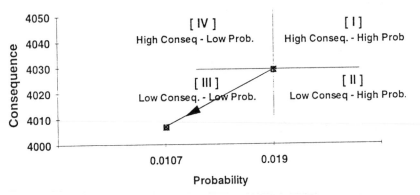

Figure 10-10. Risk reduction assessment information.

completely understand the functionality of the calculations. For most of us, it is much easier to remember a picture than a series of equations and numbers. The graphical risk assessment procedure is designed as a visual tool for a wide audience. Graphics and plots are commonplace, whereas risk is an abstract quantity. Graphing the abstract mathematical results helps a large audience of non-theoreticians understand the practical implications of risk, risk measurement, and risk assessment. This is the real value of the risk coordinate system structure.

People Contribution to Operational Risk

Our final component in measuring operational risk is the contribution directly traceable to people. People are the key component of every system, from its design to construction, to operation and maintenance. They are absolutely critical to its success. Unfortunately, close inspection will show us that people often contribute to systems' failures. Any of these failures are reflected in the equipment and product risks we've already quantified. For our purposes in measuring operational risk, however, we will look only at the most obvious costs related to people. Chapter 11 will deal extensively with human effects on risk.

Not only do people contribute to the causes of failure, they also make up some of the largest costs of failures. To calculate the people component of operational risk, you will need to tally all of the contributing costs you can find, including such things as workers' compensation, cost of liability, and even the cost of labor sitting idly. When there is a failure, nothing can be produced, yet in most cases people wait on site and are paid to do nothing until the failure is fixed. This usually even includes contract labor. This may be a fairly small amount of money in situations that are not labor-intensive. Consider, however, the situations where hundreds or thousands of employees rely on the availability of a single system. The payroll meter is running for all of these people even when a failure has made the system unavailable. A particularly good example is business that relies on large transaction processing computer systems, such as mail-order companies, airline reservations, and financial services.

Reducing Operational Risk

With the three major contributing areas to operational risk quantified, the stage is set to plan and implement problem reductions in the areas of greatest risk. These problem areas may also be candidates for more in-depth study with RCM or Risk-CM projects. Almost as important is the license that the operational risk measurement results can give you to lessen attention on areas that were identified as low risk.

Once the initial assessment is completed, the key issue becomes feedback. This means that data that was just compiled from current routine operations and processed into risk information are fed back to the process in terms of improvements and modifications of maintenance tasks and operational procedures. This feedback activity, shown in Figure 10-11, includes three perspectives on the operating environment: the actual process under study, the data analysis and risk calculations that use the data produced by the process, and the experience-based analysis that results in improvements to risk. Each of these perspectives has many dimensions and complexities within it. They all depend on one another to improve risk, but the bottom line for change is dependent on the people responsible for the process. Statistics and applied mathematics cannot, by themselves, change the process. The analytical information produced from the risk calculations identifies specific areas of high and low risks. This is the limit of utility of the numbers. The risk-based improvements are determined by the world's leading experts on operation and maintenance of the plant's processes. These experts are the people who, on a day-to-day, year-by-year basis manage and operate the equipment and systems.

Let's look at an example of this feedback cycle. In this case, the failure data analysis indicates that seal failures on certain pumps are the highest risk failure modes, followed by motor misalignment, and then bearing wearout. This information supplies irrefutable evidence of the priority in which specific problems must be addressed. Generally, the more well-

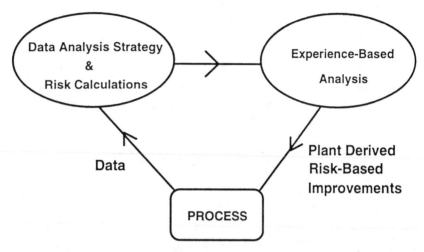

Figure 10-11. Data—risk information—process improvements feedback activity.

defined the problem, the more well-defined the solution. In this case, as seals and motor problems are resolved, the failure data resulting from the process and subsequent risk calculations will produce a new priority ranking of risks to be improved. Then the problem-solution procedure starts again. In other words, the reduction or elimination of one set of problems automatically promotes a set of risks into the high-priority position. The cyclical feedback of process performance to identify high- and low-risk failure effects, subsystems, and systems provides a solid foundation for continuous improvement.

Summary

Operational risk connects *corporate risk management* to *plant operations and engineering* by providing a common measurement framework. The measurement and assessment of risk is performed using available plant data that represents an exact representation of history. The limitation of this procedure is that it considers actual occurrences and not what *can* occur. The risk assessment methodology indicates that a company whose risk is being modified by reducing *both* consequences and frequencies is a better risk than a company that has the same overall risk reduction but the effective direction of reduction is toward higher consequences at lower frequencies.

CHAPTER 11

People: The Newest Measure In Risk

The control man has secured over nature has far outrun his control over himself.

Ernest Jones (educator, radiation physicist)

People are the main assets of any business, and at the same time, its main liabilities. There would be no plant, product, or service without the actions of people. People are an integral part of plant operation and maintenance. This fact is indisputable. What is not clear is exactly how people contribute to the safety and risk of plant operation. This is a very subtle area, and risks here are the most difficult to quantify. This area of research is a growing part of human reliability analysis studying the relationships between worker schedules, the natural biological rhythms of the human body, and worker performance [1].

The Human Contributions To Risk

In discussing human contributions to risk, it is useful to distinguish between two kinds of errors:

1. "Active errors" result in almost instantly observable effects. Generally, active errors are associated with direct, responsive operations, such as those performed by air traffic controllers, pilots, and to some extent, process control operators. These people, and the systems in which they operate, detect errors, and feed back information directly to them.
2. "Latent errors" have consequences that are not expressed or realized for a relatively long time. Latent errors are not observed until they combine with other factors. Such errors are most likely to arise with managerial personnel, designers, construction workers, and maintenance personnel [2].

Chapter 5 defined a system as ". . . a collection of things that together perform one or more functions." Even though this term is commonly used in reference to physical or engineering-based components, an integral part of any industrial operation lies with the people who operate and maintain the equipment [3]. In two independent studies of operational downtime performed on nuclear power downtime events in the late 1970s and early 1980s, the single largest root cause category was human performance. The other categories were component failures, design deficiencies, manufacturing, external, and documentation [4–5]. The actual percentages of downtime events that fell into the human performance category was 42% in one study and 52% in the other. Regardless of which of these values you choose to believe, the number of operational failures fundamentally caused by human performance was very high. At the time of the writing of this book, there is no substantial database across industries that assesses human-performance-related errors.

It is fairly well-known in insurance underwriting that humans are the primary contributors to operational risk. Personnel errors, especially latent errors, are a current topic of considerable concern. Some experts believe that while our technology is increasing equipment reliability, it is actually reducing the human reliability of its operation [6]. Find this surprising?

There are at least two factors that support this premise. First, machine or process operation used to be a more direct "hands on" activity. As process designs increased in complexity and size, computer automation has promoted people to higher level, less "hands-on" tasks, far removed from the process. Control has been made more precise by removing local human intervention and placing humans in a remote control room full of computer displays. The information operators receive is channeled through computer interfaces and displayed in color, touch-controlled video screens. Systems have defenses against the failures the designers knew about. They are usually defenseless against the rest. Thus, for an accident to occur, a sequence of highly unlikely events must occur in the right order at the right time. Latent failures generally are major players in these events. It has been demonstrated in many accidents that the technology deluges operators with information they don't want and inhibits them from obtaining the information they need to know. Technology applied to process management operation and maintenance does not, and cannot, contain all of the required human checks and balances for active and latent error detection and correction. As a result, catastrophic events, however unlikely, cannot be eliminated from risk management.

The second factor is related to the biological or circadian clock present in people. This "clock" is more than a figure of speech. In humans and other animals it is a small cluster of cells about 0.3 mm in diameter known as the suprachiasmatic nucleus (SCN). It is located just above where the broad optic nerve trunks cross over each other on their way from the eyes to the brain. Experimentation with the SCN has produced exactly what you would expect from a biological clock. Here are some examples.

If the SCN is destroyed in experimental animals, the normal intervals of sleep and consciousness are permanently changed. The previous patterns of sustained sleep and wake periods are replaced by many short intervals of alertness and sleep, seemingly randomly distributed throughout the twenty-four hour period. The total amount of sleep was the same except that it occurred in many small "catnaps." This behavior has been also observed in humans whose SCN has been destroyed or damaged by tumors.

Scientists, studying the SCN *in vitro,* have observed that the cluster of cells continues to generate a regular circadian pattern of functions all by itself. Also, to some extent, biological clocks are replaceable. "Installing" an SCN from one animal into another who has had its own SCN destroyed restores the recipient's circadian rhythms with one primary difference. The second animal now has the sleep and wake patterns that were observed in the donor animal before SCN removal [7].

Fatigue or Reduced Alertness: A Symptom (Not Root Cause) of Failure

Everyone, to some extent, has felt the effects of fatigue due to lack of sleep. The risk becomes real, or using another word, non-theoretical, when people perform the duties of their jobs in a state of fatigue. It wasn't that long ago that human beings' lives were controlled by the sun: work by day, sleep by night. Now we are taking this same species and expecting them to control and maintain complex technological processes routinely in a safe, cost-effective manner on a 24-hour schedule.

Fatigue and other human factors may appear to you as a human risk contributing factor that is well understood and designed out of existence by our technology. Nothing could be further from the truth. In fact, approximately 30% of all commercial airline accidents [8] and 80% of all general aircraft accidents [9] have pilot, crew, air controller, or maintenance personnel as a contributing probable cause factor. It is also worth noting that industrial accidents such as Three Mile Island, Bhopal, Chernobyl, and the *Exxon Valdez,* just to mention a few, all occurred in the middle of the night. It has been estimated that the U.S. national cost of lost productivity from human fatigue and reduced alertness is $70 billion dollars [10]. The estimate gives an order of magnitude indication of the tremendous losses experienced each year from the effects of insufficient rest and attempting to ignore our basic physiological needs. In continuous process industries, and as more activities shift to 24-hour schedules, technological advances can literally outperform the human operators.

The risks associated with fatigue are becoming larger from the increasing control responsibilities being placed in the hands of individual operators. Computerized control, distributed control systems, and automation provide more precise regulation of manufacturing processes under the remote, computer-filtered supervision of human operators. The interface between people and technology is an area that is just beginning to be explored. It involves complex, often subtle interrelationships and has spawned a new frontier of research called human-alertness technology [11], which involves human alertness monitoring, environmental equipment, and software modeling.

Fatigue is one factor in the people contribution to plant risk. As you might expect, there are many other human related contributors that are not as easily identified. Everyone knows that the manner in which people operate and maintain equipment has a major effect on reliability, but the difficulty is to identify the procedures and managerial practices that are at the root cause of problems.

When a failure occurs and the root cause is attributed to "human error," I suggest to you that people on the scene are often *not* the root cause of the accident. It is easy to assign blame this way. After all, they are a highly visible part of the accident sequence. Remember, though, that people are essentially human system components "controlled" by regulatory, managerial, design, circadian/biological, and several other *real* root cause candidates. To reduce the human contributions to risk, you must consider these real root causes in the design of procedures that manage the actions of the people on the scene.

Human errors, especially latent errors, go unnoticed and are not considered important until something bad happens. Also, in some cases, the standard operating and testing procedures themselves are the major causes of failure. For example, in a recent study of emergency diesel generators for a research facility, 22 failures were observed over an 8-year period. An analysis of failure events determined that 10 of them had as a root cause the procedures used to test and operate the equipment. It is in the identification of this type of information that can save downtime and reduce the consequences when a failure does occur.

How do you adequately study latent and active errors? Clearly, if plant or corporate management are cognizant of unsafe or high-consequence practices, corrections are made. Well-defined problems yield well-defined solutions. The trouble is that accidents are caused by an interlocking web of mostly latent errors. Each latent failure by itself is insufficient to cause the accident. The errors must occur in the right order at the right time. A pessimistic view of this is called fate. Some people view these events as random. My view is that improper or insufficient measurement to identify faulty procedures allows some industrial accidents to occur.

To refine or develop the measurement plan, let's first investigate the type of errors involved in major accidents. An analysis of several major accidents has identified classifications of "error origins" or "root causes." In case you are not familiar with the events, here's a brief description of each.

Three Mile Island: 4:00 a.m. March 28, 1979 [12]

Due to the introduction of water into the instrument air system, the turbine tripped and feed water pumps shut down. Operators failed to recognize that a relief valve was stuck open, which allowed some of the primary coolant to escape into the containment area. High-pressure injection of water was reduced by the crew, causing even more residual heat buildup increasing reactor core damage.

Bhopal: 12:15 a.m. December 3, 1984 [13]

Methyl isocyanate gas, used as an intermediate product in the manufacture of a pesticide, was accidentally released from a plant, killing at least 2,500 people and injuring about 10 times this number. It is the worst incident that has occurred in the history of the chemical industry.

Chernobyl: 1:24 a.m. April 26, 1986 [14]

Engineers systematically removed all safety controls to test whether or not the "coast-down" capacity of a turbine generator would be sufficient, given an appropriate voltage generator, to power the emergency core cooling system for a few minutes. This would fill the time it took to get the diesel standby generators into operation. The reactor was known to be extremely unstable at power settings below 20%. The test engineers managed to stabilize the power setting at 7% with only a small number (6 to 8) of control rods still in the core. At 1:24 a.m. reactor stability was lost, the neutron chain reaction became what is known as "prompt critical," and the subsequent explosion caused a major airborne release of radioactive material. Over 300 people were killed, 1.5 million others contaminated, and hundreds of square miles of farmland made unsafe.

Herald of Free Enterprise: 6:27 p.m. March 9, 1987 [15]

The *Herald of Free Enterprise* ferry departed Zeebrugge's dock with its inner and outer bow doors still open. As the ferry increased forward speed, water came over the bow sill, flooding the lower deck. Within two minutes the ferry capsized in shallow water, coming to rest with its starboard side above water. At least 150 passengers and 38 crew lost their lives.

Challenger Space Shuttle Explosion: 11:38 a.m. January 28, 1986 [16]

An O-ring seal failed on one of the external booster tanks 73 seconds after liftoff. The hot gases escaping through the leak pierced the main fuel storage area, causing an explosion that destroyed the entire rocket vehicle and killing all seven astronauts. The resulting investigations identified several managerial and procedural flaws.

King's Cross Underground Fire: 7:25 p.m. November 18, 1987 [17]

It is postulated that a discarded smoker's cigar or cigarette ignited grease in general rubbish under the track of the No. 4 wooden escalator leading out of the subway station. Once the fire was located by the underground authorities, incoming trains were ordered not to stop at the station. This order was not received by the train drivers, and trains continued to stop during the fire, off-loading passengers into the underground station. People in the station had difficulty finding suitable, unlocked, or unblocked exits. The station itself did not have an evacuation plan. The fire spread very quickly, killing 31 people, severely injuring several others.

Clapham Junction Rail Collision: 8:13 a.m. December 12, 1988 [18]

During rush hour commuting, a train crashed into the rear of another stationary train on the same track. A defective signal indicated that the track was clear. A third train then crashed into the wreckage that sprawled across neighboring tracks. A fourth train would have become involved in the accident if debris had not short-circuited the live rail. The accident killed 35 people and injured over 100 more. Railway union official indicated that signals and telecommunications staff were working 60 hours a week and had sometimes been forced to work by torch light because of insufficient staffing.

Flixborough Explosion: 4:53 p.m. June 1, 1974 [19]

The Flixborough Works of Nypro (UK) Limited were virtually destroyed by a chemical explosion. The blast originated within one of the chemical reactors in the facility. Of the employees working in the site, 28 were killed and 36 others seriously injured. The casualties would have been more severe if the explosion had occurred during the daytime hours of the normal work. Outside the plant, there was widespread damage and some injuries but no deaths. Property damage estimates indicated at least 1,821 houses and 167 shops and factories had suffered some damage, ranging from severe to minor, depending on their proximity to the plant.

Camelford Water Pollution: [Exact time unknown] July 8, 1988 [20]

Thousands of people became ill with stomach aches, diarrhea, nausea, ulcers, and other symptoms, after 20 tons of aluminum sulfate were accidentally injected into the drinking water. While there were no direct fatalities, it was estimated that at least 500 people would suffer permanent, long-term, serious medical problems such as, ulcers, arthritis, and kidney problems. Over 60,000 fish were also killed when the poison was flushed into neighboring rivers and streams. While the plant was unmanned, a substitute delivery driver mistakenly unloaded the chemical into a process tank instead of a storage tank, from which it quickly entered the public water supply. The error was unnoticed until serious complaints caused the water quality to be checked.

Piper Alpha Explosion: 9:31 p.m. July 7, 1988 [21]

A high-pressure, natural gas leak caused an explosion and subsequent fire on the *Piper Alpha* oil platform in the North Sea, 120 miles off the coast of Scotland. Out of the 227 men on board the platform, 167 of the crew and two rescuers died. Two thirds of the platform melted from the intense heat and disappeared into the sea.

Table 11-1 lists the error origin results compiled from one study [22]. Notice that several errors together caused each of the accidents. This point can be seen quite clearly from the case study reports themselves. A component failure, cognitive error, or substandard human performance does not, by itself, precipitate severe accidents. They are caused by the fortunately rare combination of many errors.

Also in Table 11-1, it is apparent that most root causes are managerial and procedural in nature. This means that equipment and process system failures are not solely the cause of major accidents. Accidents are caused by a special sequence of events and involve practices that are routinely

Table 11-1
Root Cause Analysis Summary of Major Accidents

Accident	Gov't & Regulatory	Procedural & Management	Design	Maintenance	Operator	Training
3 Mile Island	3	7	2	1	0	2
Bhopal	3	26	6	4	4	0
Chernobyl	1	5	3	0	2	0
Free Enterprise	1	8	2	1	0	0
King's Cross	1	6	0	1	0	0
Totals:	9	52	13	8	6	2

occurring. In other words, the practices that can, in part, cause, support, or reinforce an accident event are in operation today. The problems are how to identify them and how to correct them.

Another study looked at human factors associated with a group of major accidents using a methodology specially developed to identify the nature of the management and organizational failures and their root causes [23]. Table 11-2 is taken directly from this study and lists the results in a very detailed manner. It categorizes each accident with regards to management failures that occurred and the root causes.

Notice in this table there are several management failures, but only three root causes. The nature of these root causes can be interpreted as a complex mixture of technical and human-related faults. In my opinion, it is this area, the "human-centered" compared to the "reliability-centered," that is the next major frontier in the spiral of continuous improvement.

Fallible decision-making is a basic part of life. People will always make mistakes. It is idealistic to attempt to eliminate all mistakes. This is an attractive goal to talk about in corporate board rooms, but in practice, it is unachievable. The real task is to ensure that the adverse effects of poor decisions can be quickly detected and proper responsive actions are taken. Equipment failures by themselves do not cause major accidents or significant downtime events. These incidents are caused by a sequence of events that together define accident scenarios.

The next section discusses a procedure that refines the measurement strategy to detect time-based effects on system performance. The errors we are attempting to discover are subtle, latent, and most of the time by themselves, innocuous. The method is designed to measure the time dependence of operational errors for comparison with managerial, regulatory, and procedural guidelines. There is one fundamental objective for this type of measurement: to identify little problems before they become big ones.

Circadian Risk Analysis of Downtime Events

Up to now, the discussion of the human contribution of operational risk has mainly been centered on describing why this aspect of analysis is important. I hope you are convinced that people are the major contributors to risk in any facility. Now the discussion turns to addressing the tough question of measurement. How can a facility measure its "people" operational risk component and reduce it through the identification and elimination of high-risk management practices and procedures? Needless

Table 11-2
Taxonomy for Management Failures and Associated Root Causes

Root Causes	Management Failures	Scenario Where Identified
1. Lack of strategic communication system	1. No effective two way communication system	1. Challenger 2. Herald of Free Enterprise 3. Kings Cross 4. Clapham Junction
	2. Confused reporting lines	1. Challenger 2. Three Mile Island
	3. Poor information exchange	1. Kings Cross 2. Clapham Junction 3. Camelford 4. Piper Alpha
	4. Insufficient employee involvement	1. Piper Alpha
2. Lack of technical understanding	1. Expert knowledge not present	1. Challenger 2. Camelford
	2. Poor organization and planning	1. Herald of Free Enterprise
	3. Management failed to resolve technical problems	1. Flixborough
	4. Inappropriate written procedures	1. Chernobyl
3. Management structure	1. Inadequate definition of roles and responsibilities	1. Challenger 2. Herald of Free Enterprise 3. King Cross 4. Clapham Junction 5. Flixborough 6. Camelford 7. Piper Alpha
	2. Inadequate safety organization	1. Challenger 2. Herald of Free Enterprise 3. Kings Cross 4. Clapham Junction
	3. Complicated decision making process	1. Flixborough

Reprinted with permission from the American Institute of Chemical Engineers, 1993.

to say, this is very difficult. There is no general methodology in existence today that can identify and subsequently eliminate management practices and procedures that contribute to operational risk. The following case study is presented as a direct approach for doing this by using historical failure information for continuous improvement and risk reduction [24].

Case Study: Circadian Risk Analysis of Operational Downtime Events

Current technology and methods are highly successful in identifying and minimizing failures in equipment and systems. However, the technology is not inherently effective for identifying or preventing downtime events caused by the combination of hidden human failures that are inevitable in any organization. Although the term generally used for operator or maintenance error is "human error," the *root* cause often does not lie with the individual, but rather with the management practices that direct their actions. Historically, there has been no practical way of assessing these complex, latent human-related effects on operational risk, except in hindsight from accident investigations.

Recent work has revisited the connection between human reliability and equipment reliability. Recognizing that people are vital system components subject to circadian (time of day) cycles, this case study describes a method for identifying high-risk operational practices and procedures using this frame of reference. The word *circadian* is derived from the Latin words *circa* (about) and *dies* (day) and describes phenomena characterized by cycles, such as the 24-hour cycle of biological or physiological activity. The method combines "statistical identification" with "management review" of pipeline downtime events in a circadian framework. The ultimate objective is to eliminate the organizational root causes of downtime events.

This case study applies the circadian risk analysis method to a liquid petroleum products pipeline system, using two years of operational downtime data. The following questions, among others, are addressed regarding the frequency of downtime event occurrence, event cost, and event risk: "Which times of day have the highest and lowest probability, cost, and risk? Why?"

Data Description

A pipeline company operates 4,100 miles of liquid petroleum products pipeline extending from the Gulf Coast of Texas through the Midwest, with branches into the Chicago area and the Northeast United States. The types of products and quantities shipped vary on a daily basis. There is a general six-month seasonal cycle in shipment requirements. From October 1 through March 31, the pipeline generally operates at full capacity with all pump stations involved in product transport. This time interval is defined as the peak season or peak category. The off-peak category is April 1 through September 30 when pumping rates are lower and not all stations are continuously operating. Downtime event data are divided into these two categories and analyzed separately. Because the purpose of this case study is to introduce the circadian methodology, results are presented for just the off-peak category. Similar findings were obtained from the peak category.

Company-wide pipeline operation is controlled from a central facility in the corporate headquarters. Equipment downtime is automatically recorded in the control computers, and these data are downloaded regularly to a computerized maintenance management system. Each record in the downtime file corresponds to a specific downtime event in the 4,100 miles of pipeline operation. The data required to perform this study were extracted from this computer database. The downtime event data items required to apply the method are:

- Date and time of day of the event
- Failure description
- Lost barrels to measure the volume of product that is not shipped as initially scheduled. It is the measure of loss, or failure consequence of each occurrence. Lost barrels is computed as follows:

$$\text{Lost barrels} = (R_i - R_f) * \Delta t$$

where R_i = Initial pumping rate
R_f = Failure induced pumping rate
Δt = Downtime

In general, this measure of loss or failure consequence must accompany each failure occurrence. Ideally, the cost implications of pipeline failures should be computed in terms of dollars. The lost barrels measure is an approximation toward this ideal.

In practice, the barrels lost to the initial pumping schedule, the date-time, and failure description are the essential properties of each downtime event from an operational perspective. Location of the failure is not required in this data compilation because we are examining the entire pipeline in the circadian framework. This information may prove useful however, for future analysis.

Method of Analysis

Stage I: Data Preparation

All failure events are considered in this data analysis *except* downtime events that occurred for reasons beyond the control of the company. These include downtime events related to electric power losses when the utility companies failed to supply power, and shipper problems when the customer was unable to deliver product to the pipeline system on the planned schedule. Most of the time, the latter occurs because of equipment failures in the shippers' delivery systems.

As stated previously, downtime data are divided into two categories: off-peak and peak. This removes one known time dependent factor from the data.

Stage II: Loading the Circadian Matrices

For each data file, off peak and peak, each downtime event was categorized by its day of week and time of day (a.k.a. circadian location). Days of the week are identified numerically with the numbers 1 (Sunday) through 7 (Saturday). The time of day is indicated in military time.

Two circadian matrices of 7 columns and 24 rows were defined; one to accumulate downtime frequencies and the other to store the associated lost barrel consequences. Each element of the frequency matrix is identified by its row-column combination and indicates the number of downtime events that occurred during that hour on that particular day of the week. The lost-barrel matrix has a similar definition. The risk matrix is computed by multiplying each element of the frequency matrix by its corresponding element in the consequence matrix. This computes the operational risk for each time-of-day/day-of-week interval.

Stage III: The Application of Risk

Risk is defined as the product of probability and consequence. We apply an equivalent, customized form that is proportional to Equation 8-1. Event frequency is proportional to event probability, and consequences are being measured in terms of lost barrels. The definition of operational risk for this analysis is:

$$Risk = Frequency * Lost\ Barrels \qquad (11\text{-}1)$$

Figure 11-1 displays the risk distributions in the circadian planar framework (time of day—day of week). Operations are run continuously on 12-hour shifts that generally change at 6 am and 6 pm. The first half of the risk plane along the time-of-day axis corresponds to the day shift and the second half, the night shift. Figure 11-1 depicts intermediate results incorporating data censoring and the circadian categorization of the admissible downtime events.

There are three such distributions: failure frequency, failure consequence (lost barrels), and risk. We focus our attention on the risk distributions for the very specific reason that it incorporates the effects of the two stochastic functions: event frequency and event consequence.

Stage IV: Risk Ranking of Time Intervals

The risk data, shown in Figure 11-1, is analyzed by summing the time-of-day and day-of-week dimensions separately. In each case, the resulting risk ranking gives another profile on the time intervals that contribute the largest and smallest to the total operational risk. The objective is to use the downtime information in the most effective manner to identify procedural, managerial, or other root causes of downtime.

Profile #1: Time-of-Day/Day-of-Week

Figure 11-1 shows the relative risk in each category across the circadian plane. We first rank each [time-of-day (TOD)/day-of-week (DOW)] interval from highest to lowest risk.

Table 11-3 lists the partial results. Notice that out of the 168 (7 × 24) time intervals, *only 7% of these intervals together contribute roughly 50% of the total operational risk,* encompassing 117 of the total number of down time events. In terms of the number of downtime events, *50% of the total risk is included in only 12% of the total number of downtime events.* This information identifies specific sets of downtime events to be reviewed for the identification of any common causes of system failures.

Profile #2: Time-of-Day

The time-of-day risk distribution statistical properties are computed by using the risk contributions from the day-of-week values in the circadian planes. The results are given in Figure 11-2. The graph shows certain times of the day that appear to have higher operational risks than others.

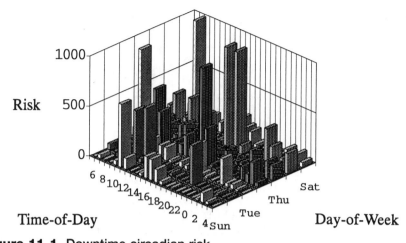

Figure 11-1. Downtime circadian risk.

Table 11-3
Time-of-Day/Day-of-Week Risk Ranking

DOW-TOD	Number of Events	% Total Risk
Sat-8	11	13
Tue-9	7	21
Fri-17	13	28
Fri-19	5	32
Wed-18	12	36
Sun-11	5	39
Mon-13	9	41
Wed-9	19	43
Mon-11	13	46
Mon-22	9	47
Fri-1	5	49
Wed-19	9	50
# of Events =	117	

There is a possible high-risk zone from 8 am to 10 am and another area of relatively high risk from 5 pm to 7 pm around the evening shift change.

The total risk associated with each time of day is obtained by summing the risk values for each of the days of the week. The time-of-day intervals

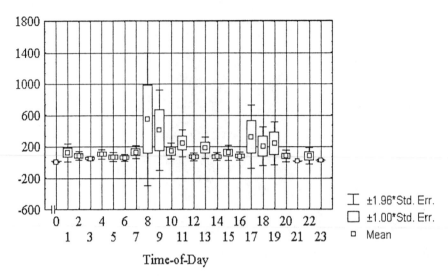

Figure 11-2. Mean risk vs. time-of-day, off-peak category.

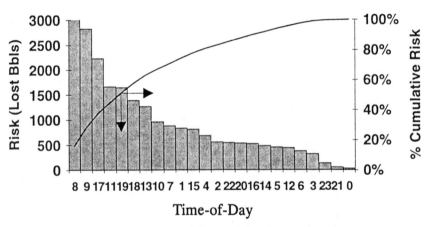

Figure 11-3. Risk ranking: Time-of-day intervals.

are ranked from highest to lowest risk contributions as shown in Figure 11-3.

Table 11-4 gives a partial numerical listing indicating the number of downtime events involved with the major risk-ranked intervals.

In this risk profile 27% of the downtime events contribute 50% of the operational risk. If time and resources do not permit adequate review of

Table 11-4
Time-of-Day Risk Ranking

TOD	Number of Events	% Total Risk
8	57	16
9	65	28
17	53	38
11	49	45
19	40	52
	264	

this relatively large number of downtime events, the risk ranking serves as a useful guide to prioritize the times of day to consider.

Profile #3: Day-of-Week

There are a considerable number of "time-of-day" factors, such as shift changes, electric power rates, and shipper schedules that make the time-of-day risk profile intuitively valuable. To observe the day-of-week risk distribution, we follow a similar procedure as with the time-of-day profile. The statistical variation of risk is plotted as a function of the day of week in Figure 11-4.

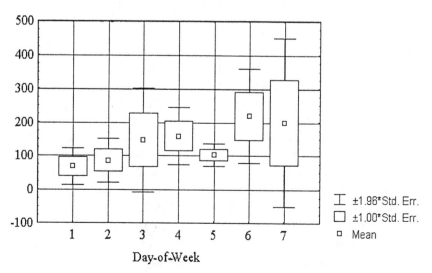

Figure 11-4. Mean risk vs. day-of-week, off-peak category.

Downtime Data Review by Risk Priority

The description of each downtime event listed in Table 11-3 and Table 11-4 was reviewed in risk priority order to determine if events were linked by common root causes such as operating procedures or other circadian based effects. While it appears that no correlation exists between these events and a specific day-of-week/time-of-day, these data sets have identified certain procedural opportunities for improving pipeline operations.

One description that is prevalent is downtime caused by "low pump suction." There appears to be an opportunity to improve the operating procedures regarding the start/stop sequence of mainline units. Most of the downtime events characterized as "low pump suction" are during unit start-up. Currently, there exists no specific operating procedure for the minimum pump suction pressure to be achieved prior to unit start-up. Another cause of low pump suction appears to be inadequate product available from the origin tankage and/or product stream switching that causes fluctuation in mainline flow creating low suction conditions. Examining product scheduling and terminal operations management appears to be the most effective approach to reducing this type of low pump suction events.

The review of downtime data also uncovered opportunities for improved communications between operations and maintenance personnel. In particular, we examined downtime event descriptions that occurred during days of the week and times of the day when the maintenance staff was not normally scheduled for work, that is, 4 p.m. to 7 a.m. Monday through Friday and all day Saturday and Sunday. A certain portion of these events appears to be for routine equipment maintenance and could have been identified at an earlier stage, reported to maintenance, and entered into the normal maintenance schedule. One circadian factor that may contribute to this is the difference between the work schedules of operations and maintenance. The effect of the different schedules may explain the pattern in Figure 11-2, where operational risk is relatively high in certain time intervals at the beginning and end of the maintenance work day. The downtime data did not give any clear common causes of failures at midday. We can only speculate that a contributing factor is the perception of operations that maintenance personnel are more or less unavailable during lunch and mid-morning and mid-afternoon breaks.

Figure 11-4 also supports the contention that the difference in work schedules of operations and maintenance affects operational risk. Operational risk increases to Wednesday, decreases on Thursday, increases on

Friday, and then decreases on Saturday. We believe that the increase on Friday is due to operations' knowledge that maintenance personnel will not be working over the weekend. The general rise in risk to Wednesday cannot be explained and may be due to the average time duration of work orders.

Case Study Conclusions

Root cause identification and specific improvements identified by the circadian analysis in this study are:

Root Cause: Lack of Technical Understanding

1. Review written operational procedures for pump start-up need to better control pump suction pressures.
2. Improve product scheduling from tankage or product stream switching to reduce pipeline pressure fluctuations.

Root Cause: Lack of Strategic Communication System

1. Improve communications between product scheduling department and field operations.

Root Cause: Management Structure

1. Recognize the inherent differences between the circadian work cycles of operations and maintenance.
   - Improve the proactive reporting of maintenance that can be scheduled as routine.
   - Improve failure reporting to eliminate operational risk time dependence on the maintenance work schedule.
   - No fatigue problems, operator errors (active or latent), or maintenance errors were identified by the circadian analysis. No adverse effects from 12-hour operational shifts were observed.
   - Risk is a high cost-benefit measure of importance for identifying managerial or procedure deficiencies. For example, we found that 50% of the operational risk is contained with only 12% of the total number of downtime events, and only 8% of the circadian elements (time of day—day of week) contained 50% of the operational risk.
   - Statistics and the risk analysis are decision *support* tools. The real identification of improvement opportunities comes from the analyst with the direct pipeline experience, not from the numbers. There is a difference between statistical significance and practical significance.

- The circadian framework is a valuable platform for developing insights into the root cause of failures. Managerial or procedural deficiencies have been hard to identify and changes are often implemented based on subjective evaluations. The circadian methodology introduces a rationale or strategy to support what changes, if any, should be made.
- The case study uncovered that the variables, day of week and time of day are, by themselves, an incomplete set to represent circadian influences. The time of day is clearly a factor that affects all personnel, including maintenance, operations, and management. However, the time of day does not affect them all uniformly. Maintenance and management generally work a single-day shift, on a Monday through Friday (or Saturday) schedule. However, operations personnel work on 12-hour continuous shifts, with daily frequencies such as 4 days on and 4 days off. Their daily circadian working schedule is independent of the day of week. Thus, to attempt to identify operator-related procedural deficiencies including fatigue, the *time and day of shift,* not the time and day of week, are the applicable circadian scales.
- Another interesting issue is the potential time lag between the error and the actual system failure. The potential for these time delays must be considered in the review of high risk downtime events. Human alertness and fatigue-related failures usually have certain signatures that can be recognized by people intimately familiar with the equipment, systems, and corporate culture.

This case study, like most other investigations, answers some questions and has raised others. The main conclusion that can be drawn from the entire study is that a circadian-based review of failure data can lead to procedural improvements correcting the latent deficiencies that are the root causes of many accidents.

Summary

The people component of risk assessment represents a new and exciting challenge. By managing employees' circadian influences and identifying high-risk procedures and management practices, the people component of risk can be greatly reduced. The next frontier in design and analysis emphasis is toward human-centered management. Reducing risk contributions associated with people is spawning new research fields and new technologies. Reducing the *people* contribution to risk is a virtual gold mine of opportunity.

References

1. Kroemer, K. H. E. "Work Schedules and Body Rhythms," IIE Transactions, Vol. 24, No. 1, March 1992, pp. 26–38.
2. Rasmussen, J., and Pederson, O. M., "Human Factors in Probabilistic Analysis and Risk Management," *Operational Safety of Nuclear Power Plants* (Vol. 1), Vienna, International Atomic Energy Agency, 1984.
3. Deming, W. E., *Out of Crisis,* MIT CAES, 1991, p. 366.
4. Rasmussen, J., "What Can Be Learned From Human Error Reports? In K. 4 Duncan, M., Grunegerg and D. Wallis (eds.), *Changes in Working Life,* London: Wiley, 1980.
5. INPO *An Analysis of Root Causes in 1983 Significant Events Reports,"* INPO 84-027, Atlanta, Ga., Institute of Nuclear Power Operations, 1984.
6. Sheridan, P. J., "Coping With Your Body Clock," *Occupational Hazards,* November 1991, pp. 47–50.
7. Moore-Ede, M., *The Twenty-Four Hour Society,* Addison-Wesley, New York, 1992, pp. 30–31.
8. "Annual Review of Aircraft Accident Data—U.S. Air Carrier Operations: Calendar Year 1989," National Transportation Safety Board, PB93-182905, p. 18.
9. "Annual Review of Aircraft Accident Data—U.S. General Aviation: Calendar Year 1989," National Transportation Safety Board, PB93-160687, p. 29.
10. Wooten, J. P., "The National Sleep Deficit: Are We Snoozing into Disaster?" *Professional Safety,* August, 1991, p. 17.
11. Moore-Ede, M. C., "Alert at the Switch," *Technology Review,* October, 1993, pp. 53–59
12. Rogovin, M., *Report of the President's Commission on the Accident at Three Mile Island,* Washington, D.C., Government Printing Office, 1979.
13. Union Carbide, *Bhopal Methyl Isocyanate Incident Investigation Team Report,* Danbury, CT. Union Carbide Corporation, March 1985.
14. USSR State Committee on the Utilization of Atomic Energy, *The Accident at the Chernobyl Nuclear Power Plant and Its Consequences,* Information compiled for the IAEA Experts' Meeting, August 25–29, 1986, Vienna, IAEA, 1986.
15. *New Scientist,* May 7, 1987, p.19.
16. *Report of the Presidential Commission on the Space Shuttle* Challenger *Accident,* Washington, DC, Government Printing Agency, 1986.
17. Moodie, K., "The King's Cross Fire: Damage Assessment and Overview of the Technical Investigation," *Fire Safety Journal,* 18 (1992) pp. 13–33.
18. *Financial Times,* December 17, 1988, p. 22.
19. Lees, F., *Loss Prevention in the Process Industries,* 1989 Butterworth, Vol. 2, p. 863.

20. *Surveyor,* (PUBWOS) August 17, 1988, p. 2.
21. *Time Magazine,* July 18, 1988, p. 36.
22. Reason, James, *Human Error,* Cambridge University Press, New York, 1990 pp. 251–257.
23. Smith, A. J., Penington, J., Lydiate, C. W., and Jackson, A. R. G.,"Human Factors in Management and Organization," International Conference on Hazard Identification and Risk Analysis Human Factors and Human Reliability in Process Safety, January 1992, Copyright: American Institute of Chemical Engineers, New York, 1992, pp. 371–388.
24. Jones, R. B., Daileda, G.,"Time of Day and Risk of Downtime Events," Second International Conference on Improving Reliability in Petroleum Refineries and Chemical and Natural Gas Plants, Houston, Texas, Copyright: Gulf Publishing Company, November 1993.

Index